THE LEGACY OF TETHYS

MONOGRAPHIAE BIOLOGICAE

VOLUME 63

Series Editors

H. J. Dumont & M. J. A. Werger

The Legacy of Tethys

An Aquatic Biogeography of the Levant

by

F. D. POR

in collaboration with

CH. DIMENTMAN

Kluwer Academic Publishers
Dordrecht / Boston / London

Library of Congress Cataloging in Publication Data

```
Por, Francis Dov, 1927-
   The legacy of Tethys : an aquatic biogeography of the Levant /
F.D. Port with the assistance of Ch. Dimentman.
      p.   cm. -- (Monographiae biologicae ; v. 63)
   Bibliography: p.
   Includes index.

   1. Aquatic biology--Middle East.  2. Biogeography--Middle East.
I. Dimentman, Ch. (Chanan)  II. Title.  III. Series.
QP1.P37 vol. 63
[QH193.M5]
574 s--dc19
[574.956]                                                  89-2481
```

ISBN-13: 978-94-010-6911-3   e-ISBN-13: 978-94-009-0937-3
DOI: 10.1007/978-94-009-0937-3

Published by Kluwer Academic Publishers,
P.O. Box 17, 3300 AA Dordrecht, The Netherlands.

Kluwer Academic Publishers incorporates
the publishing programmes of
D. Reidel, Martinus Nijhoff, Dr W. Junk and MTP Press.

Sold and distributed in the U.S.A. and Canada
by Kluwer Academic Publishers,
101 Philip Drive, Norwell, MA 02061, U.S.A.

In all other countries, sold and distributed
by Kluwer Academic Publishers Group,
P.O. Box 322, 3300 AH Dordrecht, The Netherlands.

Printed on acid-free paper

All Rights Reserved
© 1989 by Kluwer Academic Publishers
Softcover reprint of the hardcover 1st edition 1989

No part of the material protected by this copyright notice may be reproduced or utilized
in any form or by any means, electronic or mechanical, including photocopying, recording,
or by any information storage and retrieval system, without written permission from the
copyright owner.

Table of Contents

Preface		vii
A Preamble on Geographical Limits		ix
1. Early Formative History		
	1.1. What happened to the Tethys Sea?	1
	1.2. Putative Miocenic relics in the continental waters	5
	1.3. The Messinian salinity crisis	7
	1.4. The Pliocenic normalization event	12
2. The Pleistocene		
	2.1. The impact of the Plio-Pleistocenic tectonics	17
	2.2. A fresh look at the Glacial chronology	20
	2.3. The model of the last Glaciation applied to Levantine marine biogeography	24
	2.4. Pluvials, Interpluvials and lake levels	30
	2.5. Shifting rivers and captured headwaters	37
3. Eastern Mediterranean		
	3.1. The Levant Basin	41
	3.2. Some oceanographic conditions	42
	3.3. Depauperation of the zoobenthos	48
	3.4. Levantine zooplankton	52
	3.5. Positive features of the Levantine biota	53
	3.6. The progress of the Lessepsian migrants	55
	3.7. The impact of the Aswan high dam	62
4. Northern Red Sea and the Gulfs		
	4.1. The uniqueness of the Red Sea	65
	4.2. Northern Red Sea and its metahaline past	67
	4.3. Two Gulfs – two non-identical twins	71
	4.4. Origin of biota and biotic provinciality	73
	4.5. The problem of the Red Sea endemism	80
	4.6. A warmwater deep-sea fauna	83
	4.7. The paradox of the Gulf of Aqaba reefs	85
5. Halmyric Environments		
	5.1. Halmyrology – hydrobiology of waters with changing salinities	91

	5.2.	Anchialine environments of the Red Sea	91
	5.3.	Metahaline lagoons of the Red Sea	97
	5.4.	The Bitter Lake of the Isthmus of Suez	98
	5.5.	Halmyric lagoons of the Nile Delta	100
	5.6.	Residual brackish estuaries	103

6. The Continental Waters

6.1.	Limnology in the Levant	109
6.2.	The salty waters of the Jordan Valley	112
6.3.	The Dead Sea or nearly so	114
6.4.	Saline oasis springs	118
6.5.	The Lower Jordan	121
6.6.	Lake Kinneret – the Sea of Galilee	122
6.7.	The Mesopotamian primary freshwater fauna	126
6.8.	The separation of the river basins	129
6.9.	River Orontes and its complex history	133
6.10.	The Ethiopian connection	137
6.11.	The Palearctic influx and its limitations	141
6.12.	The Lebanese rivers	143
6.13.	Lake Hula and the headwaters of the Jordan by Ch. Dimentman and F.D. Por	148
6.14.	The limits of the Palearctic advance	153
6.15.	The ephemerous waters of Israel and the Levant by Ch. Dimentman and F.D. Por	156
6.16.	Polluted and manmade waterbodies by Ch. Dimentman and F.D. Por	159

Postface: The Legacy of Tethys – Or the Guise of a Conclusion	165
Bibliography	169
Geographical and Subject Index	193
Authors Index	201
Taxonomic Index	207

Preface

This book is an attempt to present a comprehensive view on the aquatic biogeography of a small but very dynamic and complex area of the globe. Luckily, this area, called here the Levant, has attracted much interest in the past and is being increasingly studied in the present. The interphasing between the knowledge of the historical and formative processes and that of the recent distributional aspects is fairly good. The recent years saw also a widening effort which expressed itself in several symposia and monographic books. Therefore I considered it possible to treat the whole subject of the aquatic biogeography of the Levant singlehandedly rather than in the presently widespread manner of an edited book. I am keenly conscious of the shortcommings of my approach of presenting much second-hand information. Possibly this is being compensated by the fact that this book has a more coherent structure and eventually a clearer scientific message. The effort spent in synthesizing the data from the widely different sources hopefully pays off in a presentation which is more easily comprehended by the average reader. For the suspicious reader, I would recommend to read first the closing chapter of this book in which the quintescence of this book and its message is presented in a summarizing manner.

Another aspect which rendered the writing of this book difficult was the need to treat the biogeographic subject on three levels, all of them necessary: the historical level, the level of the present distributional patterns and also the brief presentation of several waterbodies or environments typical of the Levant. This led to some unavoidable repetition or to extensive use of cross-references.

The taxonomic and distributional knowledge of the aquatic taxa is extremely unequal. Rather than treat the distributional patterns and mechanisms of the few wellknown taxa separately, I preferred to try to integrate these data into a general picture. This book will therefore not give the detailed information some of the readers might expect. I would like to believe, however, that it will supply some general hypotheses which can be of profit for future separate treatments.

On the strictly geographical level the correct and consistent use of locality names posed considerable difficulty. Names change in accordance with the often changing political regimes and boundaries. Moreover, transliteration from Arabic, Hebrew and Turkish is not generally accepted and consistent.

Since a book on biogeography cannot solve these discrepancies, they were accepted at face value. The reader will therefore have to face an unavoidable difficulty.

The use of illustrative material in the different chapters is unequal. This is in part an outcome of the fact that the scale used in the limnological chapters is much smaller than that of the paleogeographic and oceanographic chapters. The reader abroad is also less familiar with the peculiar geographical intricacies of the microcosmic Levantine world.

During all the phases in the preparation of this book, I had the privilege of the collaboration of Dr. Chanan Dimentman. Without his encyclopedic knowledge and carefully detailed informations, this book would have suffered to a considerable extent.

From the many colleagues who helped me in this undertaking, I wish to aknowledge the attentive reading and comments by Dr. H. J. Schnur, the excellent cooperation with our master map-maker Mr. Peter Grossman, the help of Dr. Scintila Almeida Prado-Por with the bibliography and indices, the manifold help by Ms. Ilana Ferber and the clerical help of Ms. Tamara Leff. Drs. D. Golani, N. Ben-Eliahu and Mr. H. Mienis are acknowledged for important informations. The preparation of this book was partly supported by the Israel Aacademy of Sciences and Humanities.

F.D. Por								Jerusalem, November 1988

A Preamble on Geographical Limits

The regional biogeographer often needs to face a hard decision: the tracing of biogeographic boundaries needs a considerable degree of pragmatism. The congruence between the area limits of the different species is approximate and while one waits for all the limits to be known and cybernetically analyzed, the picture in the field changes ....

The easiest way is to use political boundaries, on condition that the geopolitical unit is large enough and representative enough. What is to be done with countries, like those of the Middle East, where the states are miniaturesque, boundaries are twisted and alas, frequently modified?

When one deals with the marine biogeographic units, the term "neighbouring seas" is far from being satisfactory: distributional limits in the sea are even more fluid than those on land.

In the present book on aquatic biogeography we used a mixture of criteria, which materialize in the map presented below (Fig. 01).

The area we are dealing with bears the marks of the fragmentation of the Tethys Sea (see below): The closure of the Tethys which occurred in the Middle Eastern lands, created a "cul-de-sac" situation in the adjoining seas, the eastern Mediterranean, the Ponto-Caspian complex and the Red Sea. A whole sequence of atypical salinity and temperature conditions in these marine waterbodies ensued.

Secondly, in a closely following historically scenario, the lands exposed by the retreating Tethys underwent almost coevally a drastic and lasting desertification. The Middle East of today is situated in the center of the largest desert belt of the modern globe. For students of freshwater biogeography an impoverished, heavily constrained and peculiar aquatic world appeared.

Thirdly and finally, the gradual opening-up of the northward advancing Afro-Syrian Rift system created an impressive topographic back-bone crossing the old Tethys lands, dynamizing and rejuvenating the adjacent seas and imprinting an important hydrological axis on the continent. All these creative events initiated in the late Miocene to early Pliocene are still at work. Superimposed on them, the Pleistocenic glacials impressed a fluctuating reality which is specifically different from that of the Pleistocene of the higher or lower latitudes.

As a consequence, the easiest decision was that of delimiting the "limnological" boundaries of the region under discussion: the province called Levant (Por, 1975b) is defined as Asian branch of the Great Rift Valley. As a consequence we shall include all the basins which presently or at some time in the past were tributary to

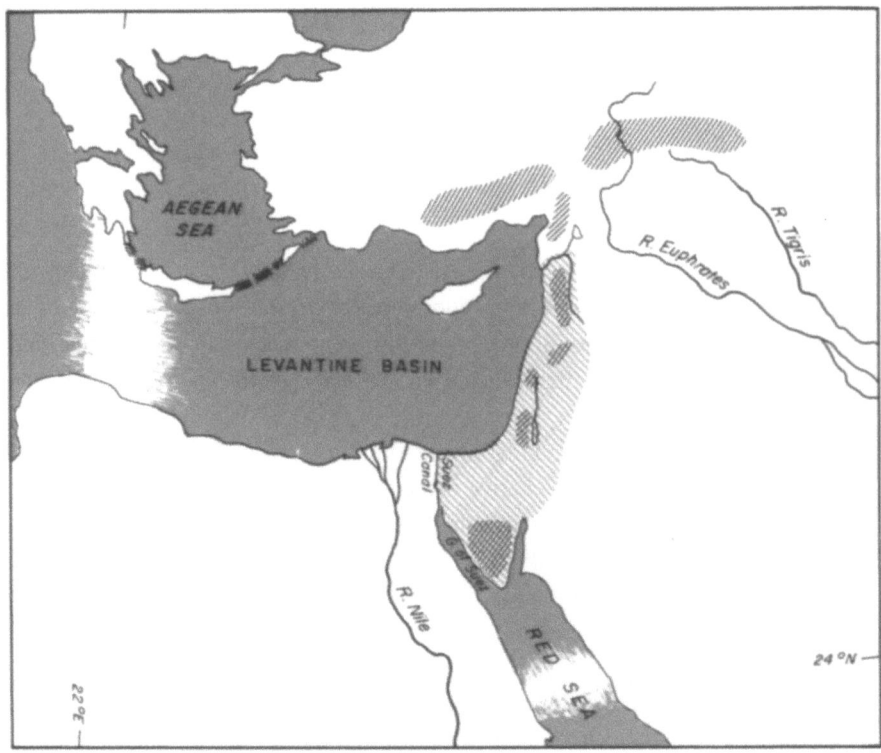

Fig. 0.1. Area map of the aquatic biogeographic province of the Levant, as used in this volume. The area of the continental waters considered in this volume is shaded. The marine areas are defined by the unprecise white bands. Mountain areas are also roughly shown (original).

the hydrological baselines of this valley. Understandably we shall deal also with the very short rivers situated between the mountains bordering the Rift Valley and the Mediterranean. There are no rivers emptying into the Red Sea, at least not at present.

The marine biogeogeographical delimitation in the adjacent seas is more difficult. There is a Levantine Basin of the eastern Mediterranean, well defined on three sides by the African, Levantine, Anatolian coast and by the island arch of the Greek islands. Towards the west, the boundary is less defined, although there are many limits proposed in literature. If a broad front connecting the Peloponessus and Crete with the north African shore will be considered, with the longitude 22° E in the middle, we have separated the most saline and warmest basin of the Mediterranean. This extreme Mediterranean "cul-de-sac" had a history of its own and is profoundly influenced at present by the massive immigration of species through the Suez Canal.

The Suez Canal therefore has to be included in our discussion and so must the Gulf of Suez, its biogeographic hinterland. The other gulf of the Red Sea, the Gulf of Aqaba or Elat, is situated on the axis of the Rift Valley and is an integral part of it. The problem again is the border towards the rest of the Red Sea. For a variety of reasons, we chose an undefined zone around the latitude 24° N. This roughly coincides with the northern Red Sea, which historically and biogeographically has much in common with the two northern gulfs. In recent literature the term Central Red Sea has gained access. Consequently the southern border of our discussion could be set further south, eventually to latitude 20° N. However, this is of little relevance in a sea like the Red Sea, where the hydrological gradients are almost ideally linear and smooth. At present, the atmospheric convergence between the southern area of the seasonally alternating monsoons and the northern area of the predominant NW winds is situated at 20° N (Edwards, 1987). But the position of this belt seemingly fluctuated recently.

We decided to include the Sinai Peninsula, first because of the fact that its northern catchments and aquatic oases are clearly Levantine. Second, the isolated mountains of southern Sinai, where freshwater fauna is residual and restricted to a few springs, connections with the north are still found. Even so, a future "Africa-centered" biogeographer will be equally entitled to see southern Sinai as a piece of Africa.

Finally, we did not include in our area either the River Nile nor the Tigris-Euphrates basin, although both had a crucial importance in the genesis of the biogeographical picture of our Levantine waters. In both these cases it was easier to set the artificial borders similar to the ones used in every regional biogeography.

We shall try to show in the following pages that there is something in common between the waters of the Tethyan heritage in our area. At the same time, we are aware that different and better definitions will be given in the future to this interesting biogeographic corner of the globe and that future points of emphasis are likely to be different from ours.

1. Early Formative History

1.1. What happened to the Tethys Sea?

The Tethys Sea, a household concept of the marine zoogeographer, was named by the Austrian geologist Suess after the wife of Okeanos. Tethys, in its classical acceptance (Ekman, 1967) was a relatively shallow warm sea which girdled the globe, more or less at the latitude of the present Mediterranean, and existed more or less undisturbed since Cambrian times. The closure of the Tethys, first in the Levant Region, afterwards at the level of Gibraltar and finally, with the establishing of the isthmus of Panama, had according to Benson et al. (1985) deep-going effects: a prevalent latitudinal warm-water world ocean has been replaced by a longitudinal ocean system, with strongly stratified temperatures. As the polar deep-water current freely penetrated into the new Atlantic Ocean, it replaced the warm deep waters of the Tethys system and eventually induced the general cooling trend of the post-Cretaceous globe.

In recent years, the concept of a circumglobal and Paleozoic Tethys ocean has suffered a dramatic restriction: once the existence of a Permo-Triassic Pangean supercontinent gained general acceptance, it became clear that the Tethys Sea rather resembled a broad, wedge-like ocean, driven into the unique continent, from east to west and ending in a cul-de-sac at the level of the present eastern Mediterranean. Laubscher and Bernoulli (1977) call this ocean the "Prototethys" and for them, the eastern Mediterranean is a relic of this ocean (Fig. 1.1). To Sengor (1979) the eastern Mediterranean is part of the Neo-Tethys, a series of "back arc" seas produced at earlier stages of the intercontinental collision in the western Paleo-Tethys area.

In Early Jurassic times both the Indian and the Atlantic oceans did not yet exist (Livermore and Smith, 1985): the western Mediterranean came into being in the Cretaceous only after the opening of the Atlantic. There are opinions which attribute to the western Mediterranean only an early Miocene or eventually an Oligocene age. While this part of the sea does not have the character of an ocean, the Levantine basin is bordered by several outcrops of ophiolites and this indicates an ocean-like genesis. Today it is clear that most of the Tethys Ocean was subducted under the Asiatic continent by the collision of India and Africa-Arabia and the only candidate which might represent the remnant of the Paleo-Tethys is the Levant Basin. While most authors speak of a late Triassic or Early Jurassic age (see Sengor 1979) Neev et al. (1985) accept a Precambrian age for this Levantine relic

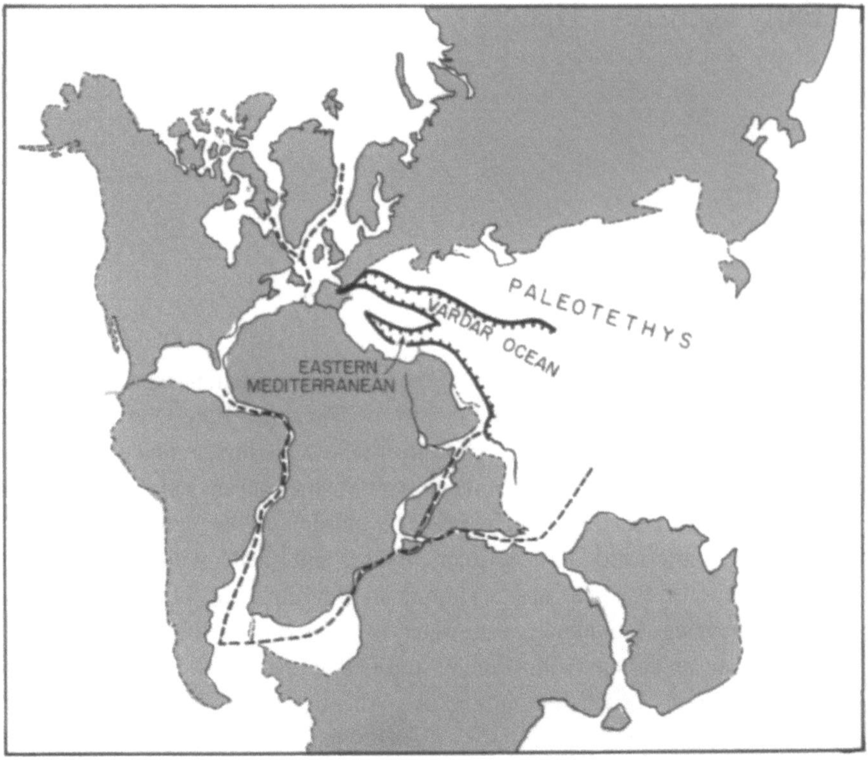

Fig. 1.1. The gradual opening of the Paleotethys across the Paleozoic Pangea (after Laubscher and Bernouilli, 1972 and Sengor, 1979).

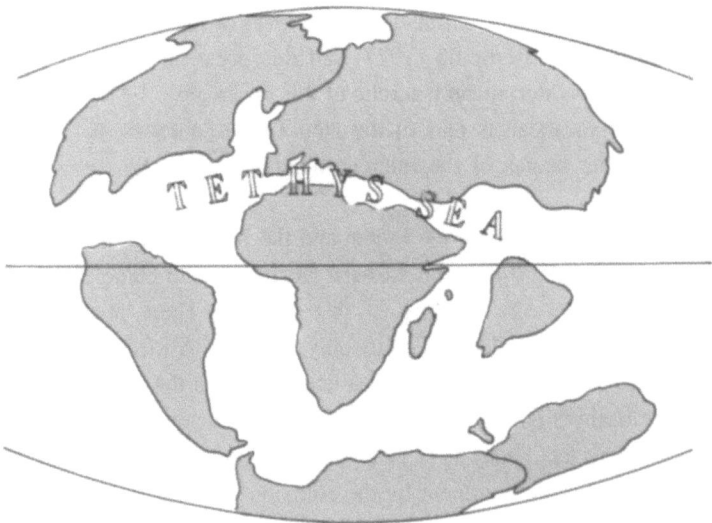

Fig. 1.2. Approximate extension and shape of the Cretaceous Tethys Sea (original, based on several sources).

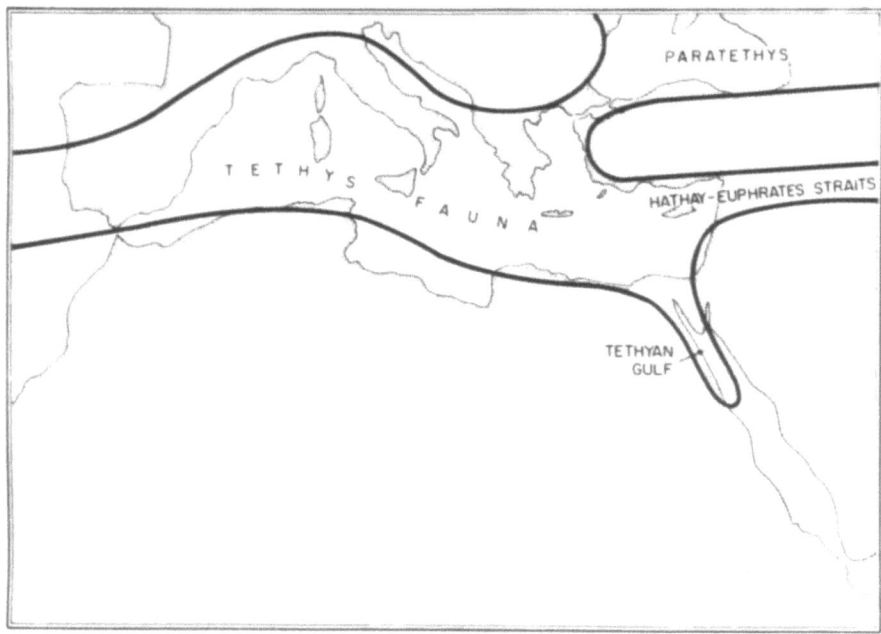

Fig. 1.3. Tethys and Paratethys during the Paleogene (from Por, 1978).

of the Tethys Ocean. Livermore and Smith (1985) are much more cautious and consider the eastern Mediterranean only as a marginal sea of the Tethys, probably one of a series of "Mediterraneans" that evolved during the Mesozoic and the Cenozoic.

It was only by Mid-Jurassic times that the tropical fauna penetrated the newly formed Central Atlantic and that the typical circumtropical Tethyan distribution pattern mentioned in the old literature came into being (Fig. 1.2).

The Levant Basin is the area where the final African-Eurasian landbridge was established in an advanced stage of the encounter between these two continental plates. Some 30 million years from now the collision between the plates will eventually wipe out the Mediterranean altogether.

At first there existed an old and perhaps the original contact between the proto-Mediterranean and the proto-Indian Ocean at the level of the "Paratethys", i.e. the Caspian-Black Sea system (the "Vardar Sea" of the geologists). The passage was subsequently closed by the upwrapping of the Iranian and Anatolian mountains. The latest active contact functioned through the northern Levant, where the Euphrates–Hatay straits were opened (Fig. 1.3).

The final closure of the Hatay–Euphrates strait, in the Jezireh area, happened in the Burdigalian epoch (Early Miocene), or according to Gwirtzman and Buchbinder (1977), in Badenian times, i.e. 13–16 million years ago. The Mid-Miocene Fars evaporites of the Persian Gulf area extend over northern Syria and reach within 10 km from the present Mediterranean Coast (Sonnenfeld, 1985). They testify for the last shallow remnant of the most recent Tethyan channel.

According to Robba (1987), the old Mediterranean separated from the Indian Ocean during the Burdigalian epoch, some 20 million years ago. However, shallow two-way traffic persisted till the final closure in the Messinian. The author compares this type of connections with a Suez Canal of sorts.

We have few data about the continental (fresh) waters of the Levant, in Paleogene times. It seems that uplifting occurred only in the Oligocene (Horowitz, 1979). Till late, i.e. after the Miocene, the predominant drainage of the waters of the Levant was in an eastward direction, i.e. to the Indian Ocean. Starting in what would be the Golan Heights of today and cutting in a NE direction through Syria, there was a Miocene gulf, reaching the Jezireh Basin, the northernmost extension of the Euphrates basin: this was the Mid-Syrian Depression, which was the main drainage system of the Levant. It was delimited towards the SE by the very old Palmyra Ridge (Wolfart, 1967). Krupp (1987) considers this drainage to be the main factor responsible for the contemporary Oriental fish fauna of the Levant. The first signs of the Pliocene tectonic activity and the connected basaltic extrusions probably interrupted this important and long-lasting drainage system, of which the Euphrates is the last, though important remnant.

During the Mid-Miocene, the Mediterranean transgressed into the Levant. An important marine gulf existed in the Beer Sheva area, represented by the Ziqlag formation with its oyster banks and the presence of the foraminiferan *Borelis melo* (Derin and Reiss, 1973).

According to several authors (see Bender, 1968) the marine Miocenic transgression reached as far as the El-Azraq basin in the Wadi Sirhan depression, where it deposited *Ostrea* beds.

The Cretaceous and the Paleogene, were the highdays of the circumtropical Tethys fauna. A coral reef fauna extended from the Indian Ocean through the Mediterranean to the Caribbean. The Mediterranean had an extremely diversified fauna of scleractinian corals and of other tropical invertebrates, well documented for the mollusks and echinoderms. With the global cooling which set-in in the Oligocene, there has been a progressive impoverishment, first probably in the Atlantic, directly influenced by the cooling Antarctic waters and gradually advancing eastwards into the Mediterranean. By the Middle Miocene, there were no hermatypic corals anymore in the eastern Atlantic and towards the end of the Miocene they died out, with some interesting exception (see below), also from the Mediterranean. Thus, a tropical fauna disappeared from our area, which according to Ekman (1967) was richer in species than that of the Indian Ocean of today. This golden age of the Tethys and the tropical fauna which characterized it, was never again to return to the Mediterranean.

To illustrate the biogeographic connections of the lagoon and estuarine fauna of the Upper Cretaceous Levant, we can use the recently uncovered and analyzed fossil fauna of En Yabrud, north of Jerusalem (Haas, 1978). According to Chalifa (1985) the fossil site of En Yabrud represents a sometimes anaerobic backreef lagoon of the Tethyan shore. Important among the finds were two species of the ancestral turtle genus *Podocnemis* (Haas, 1978), which is still alive in South America. This is an important indication for the Gondwanian character of the inshore fauna of the Levant in Tethyan times. The fauna of the continental waters must have had similar biogeographic connections.

In the Oligocene, about 40 million years ago the rifting movements formed the embryo of the Red Sea and of the Gulf of Suez (Girdler, 1984). There was a reactivation of the rifting about 25 million years ago and probably by that time the Gulf of Suez and the northern Red Sea were invaded by Mediterranean waters.

Thus the so-called Clysmic Gulf (Garfunkel and Bartov, 1977) of the Proto-Mediterranean came into being. How far south this Mediterranean gulf extended is not clear yet; but its opening to the Indian Ocean was to occur only more than 10 million years later.

During the Early- and Mid-Miocene, the Clysmic Gulf was inhabited by a tropical reef fauna similar to that of the Mediterranean. In this tropical gulf, extremely thick coral limestone deposits were laid down and the successive formations (Rudeis, Kareem and Belayim) also carried a rich fauna of benthic and pelagic foraminiferans and mollusks (Scott and Govean, 1985). At least twice during the Miocene, the Gulf of Suez underwent tectonic subsidence (Garfunkel and Bartov, 1977). These movements heralded the S-N rifting in our area.

1.2. Putative Miocenic relics in the continental waters

Turning now to the fauna of the continental waters, we have to leave, as it is nearly always the case, the solid ground of the fossil documentation.

After the opening of the Clysmic Gulf in the area of the northern Red Sea of today, i.e. after Mid-Miocene times, the fluviatile contacts between the Levant and Africa were lost. They have probably never been re-established since. Even at the highest levels of the Nile, during the Pliocenic and the Pleistocenic wet phases, the barrier of the Isthmus and the Gulf of Suez and indeed of the Red Sea was not crossed eastwards by any freshwater river.

For some time, possibly until the start of the Pliocene, a contact through the area of the southern Red Sea and Arabia was still in existence; but the huge peninsula was already under arid conditions and the way northwards was at best a steeple-chase through oases and endorheic basins.

While such a contact was still practicable for "secondary" freshwater biota, endowed with euryhaline capacities, or for aquatic insects, it was not effective for the stenohaline "primary" freshwater biota. That few of the primary freshwater fauna of the Gondwanian lands survived in our desertic area is not surprising.

We are suggesting here the existence of an isolated example for such a survival: the cyprinid genus *Garra*. The whole genus has an African-southern Asiatic distribution (Steinitz, 1954). *Garra tibanica ghorensis* (Krupp, 1982, 1983) has been described from a few southeastern tributary wadi's of the Dead Sea. The nominate species lives in southern Arabia and other near relatives in East Africa. This cyprinid genus, found in small desert streams and pools, seems to be a good candidate for a Gondwanic freshwater relic.

Koch (1988) considers that the very peculiar mayfly genus *Prosopistoma*, with two endemic species in the Levant may be considered to be a Gondwanian stock, later separated by the evolving Tethys and the uplift of the Himalayas.

The case of the subterranean cyclopoid copepod *Bryocyclops absalomi* found in caves around Jerusalem (Por 1982) is more controversial: although the representatives of this genus are found in Gondwanian lands only, the possibility of a passive transport through surface agents has to be taken into consideration: species of this genus are known to live also in the wet moss of the humid tropics.

We are nevertheless convinced that careful analysis will reveal further cases of Gondwanian freshwater relics in the Levant.

Two typical crustacean genera of the phreatic environment also found in the Levant are the syncarid *Parabathynella* and the harpacticoid copepod *Parastenocaris* (Por, 1968c; 1982).

If the assumption is correct that these are remnants of an old Pangean aquatic fauna (Schminke, 1976), then their presence in the Levant must be older than the Miocenic desertification.

In the saline springs of the Dead Sea Rift Valley and those of Lake Kinneret (Lake Tiberias), representatives of several genera of crustaceans have been found over the years: the decapod prawn *Typhlocaris galilea* (Fig. 1.4), the thermosbenacean *Monodella relicta*, the isopods *Typhlocirolana steinitzi* (Fig. 1.5) and *T. reichi*, the amphipod *Bogidiella hebraea* and recently, a species of the *Metacrangonyx* group. At least the first four species can be considered as representing a marine fauna left stranded by the Mid Miocene transgression, the last time that tropical sea penetrated inland in the Levantine province (Por, 1975b).

Similar subterranean crustacean faunas left stranded during the worldwide regression of the Upper Miocene are known from the Caribbean (Stock, 1976) as well as from several areas around the Mediterranean. Previous opinions considering this subterranean salt-water fauna as being the result of the Pliocenic marine transgression (see below) (Hubault, 1937; Fryer, 1964; Por, 1975b) are probably incorrect on two grounds: First of all, the Pliocenic Mediterranean no longer contained the tropical fauna among which the ancestors of these subterranean crustaceans have to be looked for (Por, 1986) and second, it appears now that the shortlived Pliocenic transgression did not establish normal marine environments but only a system of brackish or hypersaline lagoons.

Bogidiella hebraea belongs to a genus with much wider continental distribution and probably has a more complicated distributional history. Still, together with the other above-mentioned species, it has not yet been reported from the more northern areas of the Levant which did not suffer a very intensive Miocenic transgression.

It is interesting that *Typhlocirolana* is also found in the groundwaters of the Northern Negev valley, where because of very smooth gradients of level difference

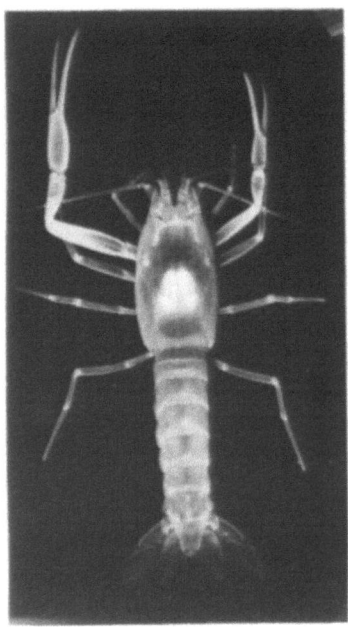

Fig. 1.4. Typhlocaris galilea Calman, considered to be a Miocenic relic in the Jordan Valley.

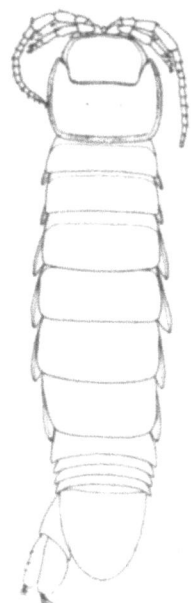

Fig. 1.5. Typhlocirolana steinitzi Strouhal, considered to be a Miocenic relic in the Jordan Valley. Uropods of right side not drawn.

between the water-tables, there is a subterranean connection between the Mediterranean and the Dead Sea phreatics (Kafri and Arad, 1978). The salinity values are also all within the brackish range (Herbst, 1982). In a sense this is a heritage of a late Neogene above-ground marine transgression into the incipient Dead Sea graben.

1.3. The Messinian salinity crisis

During the second half of the Miocene, the Mediterranean lost contact with the Indian Ocean and gradually also with the brackish Paratethys in the north. As shown by the halite and gypsum deposits of northern Syria (see above), the Levantine basin must already have had increased salinities. The Saharan desertification started by that time too, and the water budget of the semi-isolated Mediterranean probably became negative as a consequence. The Tripoli formation deposited during the Tortonian age, clearly indicates restricted oceanographic conditions (McKenzie et al. 1979–1980). See also Por and Dimentman (1985) on the first phases of this process. In the Clysmic Gulf, too, the hypersaline conditions started early: around 15–16 million years ago, after the so-called "Mid Clysmic" tectonic event, evaporite deposition began in the open waters of the Gulf (Garfunkel and Bartov, 1977).

At the start of the Messinian period, about 7 million years ago a massive and rapid glaciation event in the Antarctica, lowered the sea level by more than 40 m and isolated the Mediterranean from the Atlantic. The drying-out of the Mediterranean could have taken only several tens or a few hundred years. During 130,000 years, enormous quantities of halite were deposited in the Mediterranean basin (McKenzie and Oberhansli, 1985) representing probably 6% of the total saline content of the Atlantic. After an equally short period of deglaciation, a short-lived contact with the Atlantic led to the deposition of marine sediments in the Mediterranean. This is called the "Inter-Messinian" event by the authors. Then a second wave of sea-level lowering returned the Mediterranean to the state of an enormous salt-pan. The whole salinity crisis encompassing the Messinian period lasted for about 600,000 years. The re-establishment of a permanent connection with the Atlantic through the Gibraltar straits and the return to normal marine conditions, marked the start of the Pliocene epoch. The Messinian crisis took place between 5.57 million to 4.93 million years ago (McKenzie and Oberhansli, op. cit.).

The discovery of the Messinian drying out of the Mediterranean (Hsü et al., 1978) was the first instance in which modern historical biology was faced with the proof of a relatively recent catastrophical event of wide biogeographic significance. For the Mediterranean biogeographers, all the respected ideas had to be reconsidered. As emphasized by Por (1978), it became nearly impossible to speak of "in situ" Tethyan relics in today's Mediterranean: long before the cooling in the Pleistocene, they were wiped out by the hypersaline conditions in the Messinian lagoons.

During the last two decades of increased research related to the Messinian event, the catastrophic happening was reduced to more realistic dimensions. First, it was thought that the empty Mediterranean basin had the characteristics of an enormous Dead Sea valley, with negative altitudes of thousands of meters, corresponding to the present deep trenches of the sea. Today it is becoming clear that the basin suffered considerable tectonic subsidence after the Messinian, a subsidence which is still very active, especially in the area of the Hellenic ridge (Vanney and Gennesseaux, 1985). The Messinian crisis was not a unique event, nor separated into two phases only (see above), but interrupted or alleviated by occasional and regional influxes of Mediterranean, Paratethys or even Indian Ocean waters (for this and other aspects related to Messinian "mechanisms", see Sonnenfeld, 1985). The reopening of the contact with the Atlantic did not happen in the shape of a mighty "waterfall" through the Gibraltar portal. It rather seems that a Bosphorus-like narrow strait could have been in existence during the Messinian (Sonnenfeld and Finetti, 1985). Sonnenfeld even accepts the possibility of large-scale seepage through the dry Gibraltar portal, which could turn the Mediterranean into a big anchialine waterbody (for anchialine basins, see Por (1985) and below).

The salinities during the Messinian regression were diverse: in the areas influenced by occasional Paratethys inflow, there were freshwater "lago-mare" conditions with *Melanopsis* and *Congeria* beds of the Italian type. In the Levantine basin there were prevalent relatively shallow hypersaline sabkha-like basins. In general, there is a gradual increase in the thickness of the Messinian evaporitic strata from west to east: Balearic Sea, 1000 m; Ionian Sea, 800 m; Levantine Sea, 1800 m (Sonnenfeld and Finetti, 1985); Clysmic Gulf of Suez, 2000 m (Scott and Govean, 1985) and even 3,500 m (Ross and Schlee, 1977); Red Sea, 5000 m (Girdler, 1984).

We can reasonably assume that the same type of increasing W-E salinity gradient which exists in the present Mediterranean always existed in the post-Oligocene sea, albeit enormously steeper during the Messinian crisis and the times shortly preceding it.

In general, an actualistic approach can help greatly in order to understand the type of biota which survived the Messinian crisis in the widely fluctuating lagoons and sabkha's of the vanishing Mediterranean (Por and Dimentman, 1985). Instead of an initial feeling that practically nothing survived the crisis "in situ", the presence of low-diversity and highly euryhaline species became accepted. Besides the few species which left scattered fossils in the Messinian sediments, one can logically admit that more species, living today together with them in Mediterranean euryhaline lagoons, were also present during the Messinian.

A tentatively named *Cyprideis-Diamysis-Aphanius* assemblage was proposed (Por and Dimentman, 1985) as characterizing the Messinian biota, with several possible additional species and regional variations.

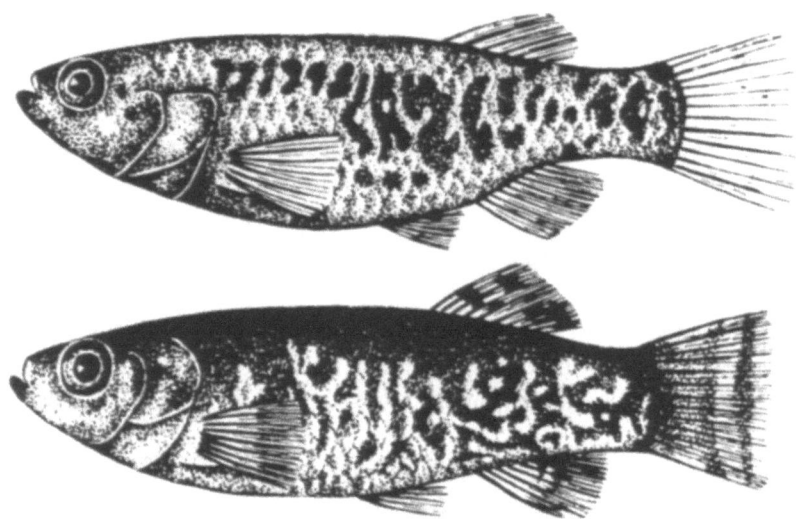

Fig. 1.6. Aphanius dispar, typical Messinian relic fish.

Table 1.1. Similarities and differences between Messinian and Paratethyan species.

Messinian species	Paratethyan species
1. Basically holeuryhaline, from brackish to hypersaline	Basically limnic euryhaline and marine hypoeuryhaline
2. Stenoionic (~sea-water composition)	Euryionic (S/CI > 2.6)
3. Shallow standing waters	Running waters, or shallow littoral
4. Moderately oxygenated waters	Highly oxygenated waters
5. Eurythermic, towards high temperatures	Eurythermic, towards low temperatures
6. Rarity of planktonic forms	Rarity of planktonic forms
7. Few hard bottom forms	Few hard bottom forms
8. Widespread taxa	High degree of endemism
9. Selection of high genetic, physiologic and phenetic polymorphism	Selection for endemic speciation in the river basins
10. Important ecological potential in high salinity lagoons of semiarid and arid shorelands	Important ecological potential in temperate river systems
11. Adaptation to drastically changing environments, and restricted possibility of retreat	Adaptation to basically stable environments, with likely possibility of up- or downstream retreat

Cyprideis littoralis, an ostracod crustacean, *Diamysis bahirensis*, a mysid crustacean, and the fish *Aphanius dispar* are found in saline lagoons of variable salinities around the East Mediterranean shores. *Cyprideis* is a characteristic Messinian fossil and fishes of genus *Aphanius* have been reported by Sorbini and Tirapelle-Rancan (1980) from the Messinian of Italy. *Aphanius dispar* (Fig. 1.6) is probably the only certain Tethyan survivor in the Mediterranean (Kosswig, 1967). But a small number of other species can be most probably added. Besides the above-mentioned species, a series of others are likely candidates, such as gobiid fishes and hydrobiid gastropods (Por and Dimentman, 1985).

A set of characteristics for this above-mentioned Messinian species complex, compared with the characteristics of the Paratethyan biota was drawn (see Table 1.1). Among the peculiarities of the Messinian biota, the fact should be emphasized that they tend to form "species groups", i.e. complexes of polymorphic populations in which the limits between the species have to be established by intersterility limits rather than through morphological differences. This is for instance the case of the *Diamysis bahirensis* species group or the group of species centering around *Aphanius dispar*.

The adaptation of the Tethyan fauna to the hypersaline conditions must have started even earlier, still in Tortonian times, in the area of the north-Syrian closure of the Tethys Sea (Por and Dimentman, 1985), but proofs for this have still to be found (Fig. 1.7).

Tortonese (1985) considered that the "paleo-endemic" species among the Mediterranean fishes of today, belong to Tethyan lineages that survived the Messinian crisis. In contrast to them are the "neo-endemic" fish species which evolved from the fauna which resettled the Mediterranean after the Pliocenic

Fig. 1.7. The Middle Eastern center of origin of the Messinian euryhaline fauna. (From Por and Dimentman, 1985). Black arrows-Pontocaspian low salinity fauna; white arrows – Messinian high salinity fauna; Bent black arrows – retreat of marine fauna.

reopening of Gibraltar. Besides the above-mentioned *Aphanius*, Tortonese also lists the fishes *Syngnathus* and *Gobius* (the three of them known as fossils from Messinian beds (Sorbini and Tirapelle-Rancan, 1980)) as well as *Callionymus pusillus*.

A very interesting case is the presence in the Messinian deposits of the western Mediterranean, of monospecific coral reefs of *Porites* (Esteban, 1979/1980). Since the reef-building corals had disappeared by that time from the Atlantic, only two possibilities for the presence of this odd coral in the Messinian Mediterranean can be considered: either survival in some isolated propitious environments or immigration from the Indian Ocean. The repeated alternance of hypersaline stromatolitic mats with the *Porites* reefs suggests that the coral species was present in the basin and could repeatedly build extensive reefs as soon as conditions improved. For the Indian Ocean contact, considered by Sonnenfeld (1985) as a distinct possibility, we see no corroborating evidence. There are no reports of reef-building corals from the Messinian Suez Gulf or from the eastern Mediterranean. Moreover, there is no indication for a pre-Pliocenic opening of the Red Sea to the Indian Ocean. Still, it is interesting to note that mono-specific reefs of *Stylophora* occur in the high-salinity El-Blayim lagoon of the western Sinai coast and that one species of *Porites* forms small reefs in the Persian Gulf (Kinsman, 1964). Vacelet (1981) considers that the Mediterranean sponge *Petrobiona massiliana* might be a "paleoendemic", especially since some subfossil material of this species was also found in Crete. Here we might have a second case of hard bottom organisms which survived the Messinian crisis "in situ". Hsü (1986) (quoted after Sorbini, 1988) recently found coral reefs in the Messinian deposits of Cyprus.

Sorbini (1988) found additional evidence for the presence of a restricted marine ichthyofauna during the Messinian of Monte Castellaro in Central Italy. In addition to the fossil *Aphanius* found by Sorbini and Tirapelle Rancan (1980) typically marine genera of fishes like *Zeus faber* and *Epinephelus sp.* were found, together with the estuarine *Atherina boyeri*. All these are Indopacific taxa, which did not survive into the Pliocene. Furthermore, Sorbini (op. cit.) reports from the Lower and Middle Pliocene, the Indopacific species *Sargocentron* cf. *rubrum* and *Hemiramphus* cf. *far*. Both these species were known in the Miocene Mediterranean basin. Since Sorbini does not believe in a Pliocenic contact between the Mediterranean and the Red Sea, these species must have survived the Messinian crisis in place. When the Suez Canal was dug in 1869, they were certainly absent from the Mediterranean; not surprisingly they were among the first species to recolonize this sea coming from the Red Sea.

Clearly, the micro-scale of the Messinian event, in itself extremely short in the geological record, is revealing a very complex picture.

1.4. The Pliocenic normalization event

Slightly less than 5 million years ago, the Atlantic transgressed again into the Mediterranean. This so-called Trubi transgression was most probably due to a massive deglaciation in the Antarctica. It is also considered to be the boundary of the Pliocene epoch, a short geological time unit, which at least in the Levantine area has an extremely complex and little-known history.

At about the same time, the old Clysmic Gulf opened to the Indian Ocean. It is a very challenging fact that the Red Sea was resettled by Indopacific biota and the Mediterranean by Atlantic ones and that no discernable mixture occurred between the two faunas. The Isthmus of Suez must have started its separating function already by the early Pliocene. It is still an open question, why the renewed tectonism of the Plio-Pleistocene, "turned" NNE along the Aqaba-Dead Sea line and became dormant along the initial NW axis, i.e. in the Gulf of Suez area. The gradual opening of the Red Sea starting from the south (Girdler, 1984) did not reactivate the "Mediterranean connection". The age of the Gulf of Aqaba is still somewhat unclear: according to Girdler (op. cit.) it was already marine by Pliocenic times. Friedman (1985), based on calculations of the sediment infill, accepts an age of 5 million years, but, based on the presence of evaporites in the southeastern area of the Gulf, he considers it to be of even older age.

The Red Sea had a tropical fauna since those times. The Mediterranean, however, was fated to a more complicated history: The Lower Pliocene or Zanclian, lasting for barely 2 million years, was a relatively warm and tectonically stable time. Nonetheless, there was no recolonization by tropical biota: everything, especially the micropaleontological data (e.g. Benson, 1975; Bizon, 1985) indicates a settling by temperate Atlantic species. Coral reefs did not reappear in the Mediterranean, not even the *Porites* reefs of the Messinian.

At least starting with the Pliocene normalization event, if not already with Miocene times, the eastern Mediterranean remained consistently an impoverished "cul-de-sac" compared with the western Mediterranean: Although temperatures were consistently higher, salinities showed wide-range and rapid fluctuations with far-going consequences for the faunal inventory. As mentioned repeatedly by the author of this book, only the invasion of the Lessepsian migrants through the Suez Canal, the first tropical marine biota to reach the Levant after the Miocene salinity crisis, led recently, and for the first time, to an enrichment of this impoverished faunal inventory.

Authors seem to agree that the Zanclian Mediterranean was still a predominantly shallow sea, and that the deep basins, especially those in the eastern Mediterranean, started to form only in the second part of the Pliocene, the Piacenzian age (see, e.g., Fabricius et al., 1985). Stanley (1985) considers that the average depth of the Pliocene Mediterranean was 1000–1500 m, compared with the average 3000 m of the recent Mediterranean.

Vergnaud Grazzini (1985), analyzing the data of the stable isotope records, reached the conclusion that during the first half of the Pliocene there was no formation of Mediterranean deep water and that the characteristics of the Mediterranean waters were those of the Atlantic.

A fairly important percentage of endemic species of the Mediterranean, for instance of the fish species, is made up by shallow water inhabitants (Tortonese, 1985). With increasing depth, the percentage of endemic species decreases (Fredj and Laubier, 1985). This might be an indication of the fact that the Atlantic

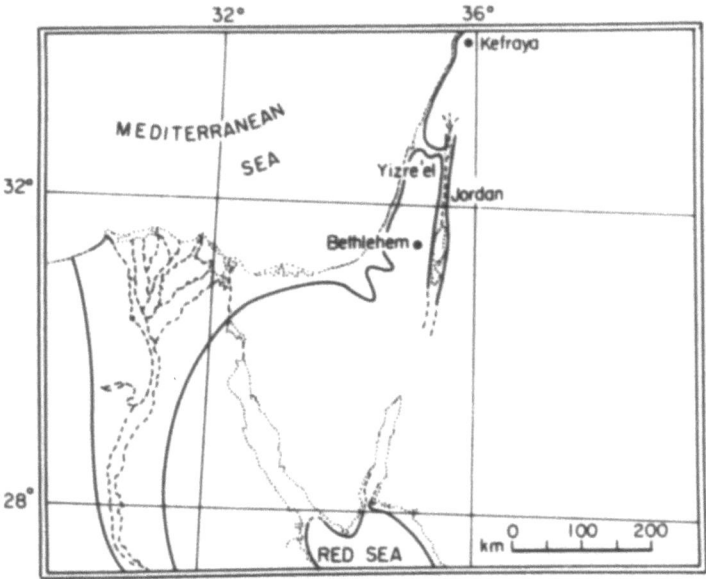

Fig. 1.8. Pliocenic marine gulfs in the Levant and north Africa (after Tchernov, 1988).

immigrants settled the deep waters much later, i.e. after the deep basins were formed. The most peculiar event of the Pliocenic normalization during Zanclian times was the marine invasion of the Nile Valley and of the embryonic Levantine Rift Valley (Fig. 1.8).

The history of the "Eonile" can be followed to Miocene times (Said, 1981). During the Messinian regression, the fluviatile errosion cut deep canyons corresponding to the deepened basement level. The bottom of the grand canyon of the Nile is 570 m below the present sea level (Williams and Williams, 1980).

The Zanclian sea drowned these canyons and a long and narrow tongue of sea water penetrated into the Nile valley as far as Aswan. The gulf was some 600 km long and its width on average only 12 km. Some drowned tributary valleys turned into fjords of up to 40 km in length (Said, op. cit.). Marine and brackish oyster beds were deposited in this narrow gulf. By Middle or Late Pliocene, the sea retreated from the Nile valley and the uplifting of the eastern African mountains began: some Pliocene lagoon sediments are found today 120 m above the present sea level (Said, 1981).

During the later part of the Pliocene, the high East African mountains in the source areas of the White and Blue Nile were uplifted and the sea retreated. The rejuvenated "Paleonile" (Said, op. cit.) re-established fluviatile conditions in the lower Nile basin: there are to our knowledge no marine or brackish relic biota testifying for the short and massive marine invasion of the Nile valley. However, a careful investigation of the subterranean waters around the Pliocenic gulf may probably result in the discovery of such species. The Zanclian transgression also brought marine waters into the Dead Sea and Jordan valley. At that time the bottom of the valleys is believed to have been at sea level (Kafri and Arad, 1978). The penetration was through the Yezreel valley though a fairly extensive embayement was also present in the plain of Beer Sheva. Horowitz (1979) dates the marine intrusion at 4.7–5.0 million years. The so-called "Bira formation" (Shulman, 1959) testifies for lagoon-like brackish conditions (see below, p. 112, 114) (Figs. 1.9 and 1.10).

In the south, in the area of the Dead Sea, the Lower Pliocene marine gulf was hypersaline and deposited the halites of the "Sedom Formation" (Neev and Emery, 1967; Horowitz, 1979). According to the latter author, after a phase of marine regression, there was a second marine transgression into the Dead Sea valley, in the Upper Pliocene Piacenzian times. It is remarkable that, despite the fact that the water-divide between the Jordan basin and the Mediterranean basin is situated today in the Yezreel Valley at only + 10 m elevation, there is no indication for a later, post-Pliocenic transgression of the sea into the Jordan valley. Most of the oligohaline and euryhaline aquatic biota which characterize the surface waters of the Jordan basin today have their origins among the Pliocene brackish invaders (Por, 1985b).

Along the Yezreel Valley and the headwaters of the rivers and swamps, there were probably also some later possibilities for a "steeple-chase" contact between

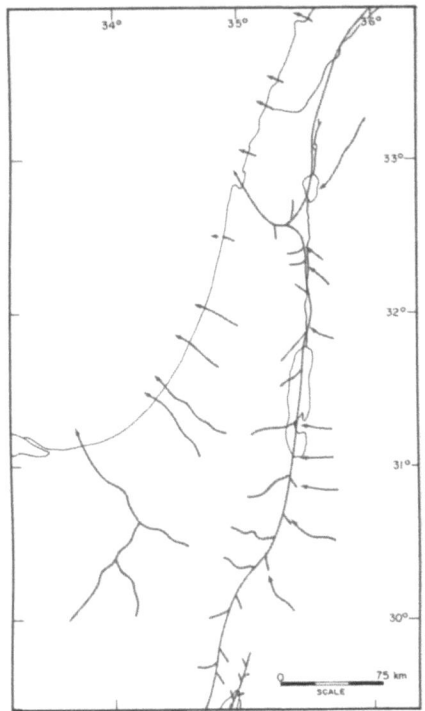
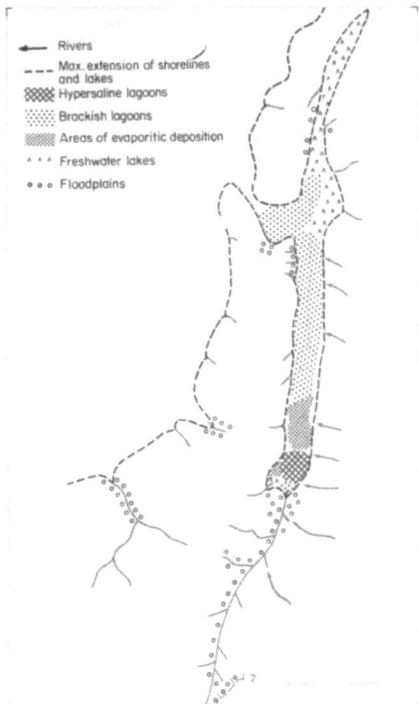

Fig. 1.9. Hypothetic map of the hydrological network during the early Pliocene in Israel (after Horowitz, 1979).

Fig. 1.10. Waterbodies in the area of Israel, from Mid-Zanclian to Mid-Piacenzian Pliocene (after Horowitz, 1979).

the Mediterranean brackish biota and those of the Jordan system (Por, 1975b). Today, the fauna of the Jordan Valley represents a well-defined sub-province where oligohaline biota predominate (Por, 1984b).

As mentioned above, we do not believe that the subterranean relic species found in the salt springs of the Jordan graben originated with the Pliocene transgression, since it seems that the "Bira" gulf did not have a normal-salinity marine fauna.

Elsewhere in the Levantine Province, the Zanclian transgression had a much more limited extension. Only in the area of Lake Amiq Golu (Lake Antiochia) and the lower river Orontes (Asi), there are marine deposits considered by Wolfart (1967) to be of Piacenzian age(?). Up to the first decades of this century, i.e. before its total drainage, Lake Amiq Golu contained fishes like *Mugil* and *Anguilla*, migrating from the open sea (Lortet, 1883). There is no doubt that there must also have been a marine-brackish invertebrate component there. Unfortunately the poorly known fauna of Amiq Golu disappeared with the drainage of this lake. Still, we might imagine it as a model for the fluviatile upstream penetration of a limited number of euryhaline marine species into the Pliocenic Rift Valley area.

2. The Pleistocene

2.1. The impact of the Plio-Pleistocenic tectonics

After the tectonic activity in the Miocene, the lower Pliocene was a relatively quiet period. Intense tectonic activity started towards the Middle of the Pliocene and it has not stopped or decreased since.

During the Piacenzian a series of dramatic subsidences and upliftings in the eastern Mediterranean began: The island of Rhodes was lifted up in the late Pliocene, while landmasses previously connecting the island to the mainland were submerged (Meulenkamp, 1985); the Aegean Basin collapsed in early Quaternary times and the Nile River sediments foundered to considerable depth, jointly with the formation of the Mediterranean ridge (Sonnenfeld, 1985). The initially shallow Levantine Sea deepened considerably even in Late Pleistocene times (Gwirtzman and Buchbinder, 1977). Similar changes also occurred in the more western parts of the Mediterranean, but it seems that the tectonic processes are still active especially along the northern margin of the Levantine Basin. This is probably also related to the rifting tectonics which characterize the Plio-Pleistocene of the Levant. If in today's Mediterranean much of the deep water circulation is generated in the Levantine Basin (Lacombe and Tchernia, 1960), this situation has to be considered as a new oceanographic development on the time scale.

When discussing the low diversity of the biota presently living in the Levantine Basin, one should bear in mind the fact that the bottom of this basin is still subject to a great deal of tectonic adjustment. Submarine landslides and turbidity currents are extremely harmful for the continuous settlement of the zoobenthos on the continental slope and the deep sea bottoms. But as we shall see in the next chapter, there were further and radical restrictions for benthic life in the Levant Basin.

The main phase of the rifting started in the S. Red Sea and gradually spread northward. According to Le Pichon and Francheteau (1978) the earliest date for the reactivation of the Red Sea rift would be around 4 million years ago. That would mean that the Indian Ocean influx into the present Red Sea basin was more recent than the 5 million years old reopening Gibraltar contact. Alternatively one may assume that the first phase of the Indian Ocean influx was due, like in the Nile valley, simply to the considerable rise of the sea-level at the Mio-Pliocene boundary.

Some authors consider the Red Sea opening to be of an even younger age, i.e. 2–3 million years (Ross and Schlee, 1977). More important still, the Isthmus of

Suez was uplifted tectonically sometime around 2.5 million years ago. It follows that from the opening of the Bab el Mandab straits and the tropical invasion from the Indian Ocean and until the isolation from the Mediterranean through the Isthmus of Suez, there might have been a short period with opportunities for a penetration of Indo-Pacific biota into the Pliocene Mediterranean. There is some fossil evidence for this (mollusks, Grecchi, 1978; forams, Wright, 1979). In addition, the presence of some "Pre-Lessepsian" Indopacific biota in the Mediterranean which existed there before the opening of the Suez Canal (Por, 1978) could perhaps be explained in this way. Such is the case of the sea grass *Halophila stipulacea* and foraminiferans associated with it and of the little pearl oyster *Pinctada radiata* (Por, 1978).

There is much more detailed information about the successive phases of the rifting in the Dead Sea, Jordan, and Syrian rift valley, called by Horowitz (1979) the "Levantine Rift Valley": Late Miocene and/or Early Pliocene tectonic movements in our area are associated with the "Erythrean faulting system" (Horowitz, op. cit.) and formed such NW-SE running depressions as the Yezreel valley in Israel and Wadi Sirhan in Jordan. At about the same time, the northern edge of the Levantine area was uplifted, interrupting the fluviatile contact with the basin of the Persian Gulf which might still have been active. According to the same author, the Lebanese and Syrian rivers of the nascent Pliocene Rift Valley drained towards the south. Horowitz (1979) suggests that they emptied to the Mediterranean through what is today the Yezreel valley in Israel.

The second phase of massive tectonic activity is placed by Horowitz (op. cit.) in the "Preglacial Pleistocene". Since the tectonics were preceded by the lava outflows known as the "Cover Basalt" (Picard, 1963), these could be magnetometrically dated (Mor and Steinitz, 1983). According to these authors and to Tchernov (1987) the impressive volcanic activity which caused this lava flows in the Levantine Rift Valley is dated around 3 million years.

What followed was the most crucial tectonic event in the geological history of the Levant: It opened the Gulf of Aqaba and deepened below sea level the Dead Sea Jordan valleys, thus creating the endorheic basin which exists there up to the present time. Further north, the rift valleys of the Ghab and of Afrin–Karasu were downwraped. But the intervening area, i.e. the present Bekaa of Lebanon is according to Picard (1970a) "something of a tectonic 'sphynx' in the analysis of the Levant graben valleys". It clearly separates the southern from the northern graben systems and stands as a geologically durable topographic barrier between them. According to Horowitz (op. cit.) basaltic flows which cover the Mettulah-Marj Ayun area separated the "southern" rift valley from the "northern" Lebanese-Syrian rift valley. From this time onwards, any interchange of aquatic fauna between the two parts of the Levantine rift had to use the occasional, quasi-accidental opportunities of the head-water capture steeple-chase. Por (1975b) considered these events as very probable and frequent, Kinzelbach (1987), though not denying their existence, emphasizes the differences between the two parts of the Levantine Rift

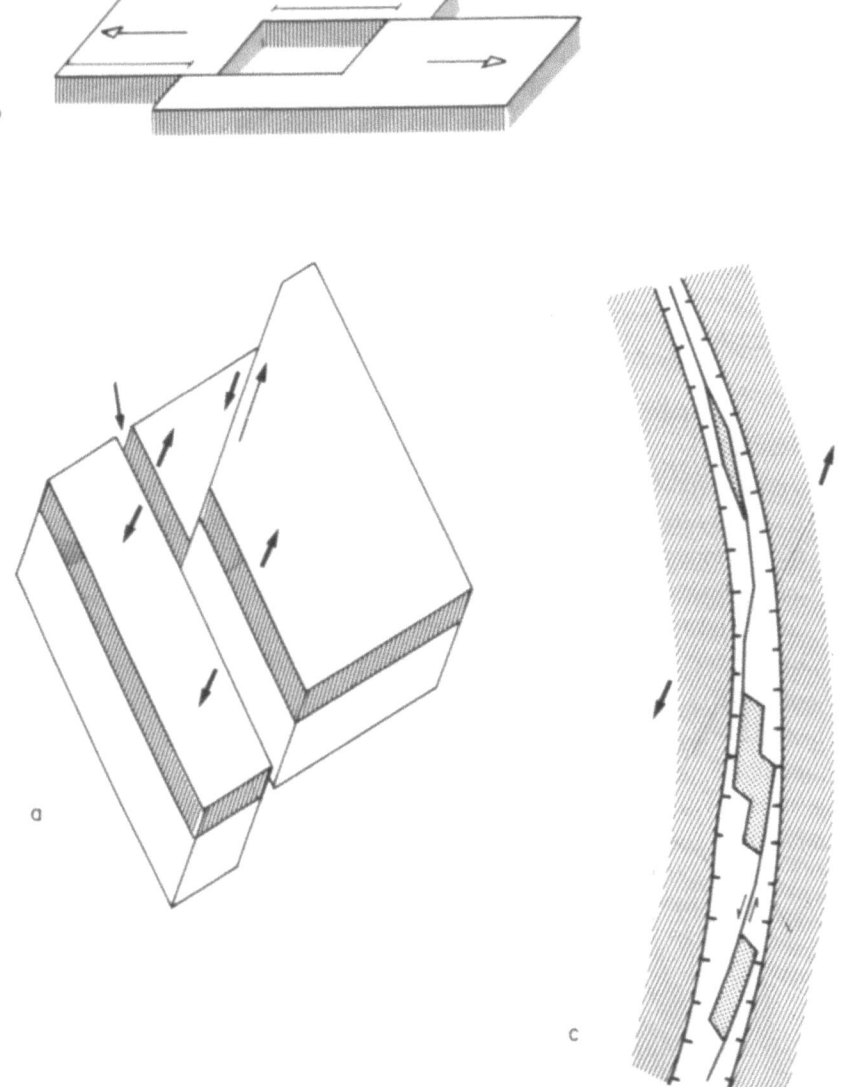

Fig. 2.1. Schematic shape of the rifting movements. a) Relative movement of the African, Sinai and Arabian plates; b) Opening of a Rhomb graben associated with the shear movement of the Levantine Rift Valley; c) Graben opening in the Dead Sea Area (from Garfunkel, 1986).

Valley, which are no doubt the result of their early Pleistocene separation. This is exemplified for instance by the subspecific degree of differenciation of the populations of the freshwater prawn *Atyaephyra desmarestii* and of the freshwater crab *Potamon potamios* (Kinzelbach, 1980). In both cases, the endorheic Jordan system and the Orontes system have morphologically different populations. Similar situations are described by Kinzelbach (op. cit.) for freshwater mollusks (see p. 127).

Krupp and Schneider (1988) see a limited function of these inter-fluviatile contacts, at least in the southward direction.

The rifting tectonics are still very active and a major phase occurred only about 18,000–20,000 years ago (see below p. 35). In the Gulf of Aqaba area the movement is a combined slip and spreading movement, while in the Dead Sea area and further north, only the slip component predominates. This movement accounts presently for an annual 9 mm of displacement (Ben Avraham, 1987) (Fig. 2.1).

2.2. A fresh look at the Glacial chronology

It is generally considered that the start of the Pleistocene coincides with the start of the so-called Quaternary Ice Age. Some authors, like Horowitz (1979) used the term "Pre-Glacial Pleistocene", but this practice is being questioned lately. Quite on the contrary, there is increasing evidence, especially from oxygen isotope data in marine sediments, that the cooling trend heralding the Ice Age already started in the Upper Pliocene, the Piacenzian epoch, i.e. about 3 million years ago. Shackelton and Opdyke (1976) consider that the start of the glaciation in the northern Hemisphere should be placed, on isotopic grounds, at about 2.5 million years ago. Nilsson (1983) accepts this date as the start of the Pleistocene.

In the Mediterranean, there are clear-cut indicator fossils for the start of the cooling: these are the "boreal guests", a set of boreal mollusks, *Arctica islandica* (Fig. 2.2), *Mya truncata, Macoma calcarea, Chlamys islandicus, Buccinum undatum*, etc. These are found in the so-called Calabrian deposits of Italy. Among

Fig. 2.2. Arctica islandica, typical Glacial visitor in the Mediterranean.

the microfossils, the foraminiferan *Hyalinea baltica* should also be added to the list of the boreal guests. Though not pertinent to the discussion at this point, it should be mentioned that none of the nordic mollusks has been reported till now from the Levant Basin. From Israel there is only the report of *Hyalinea baltica* from the earliest Pleistocene (Moshkowitz, 1968).

The appearance of this boreal fauna of the Calabrian age in the western Mediterranean was considered to be the reference date for the start of the Pleistocene on the whole globe. Recently, using the dating possibilities offered by the magnetometric method, the Calabrian was calculated to be roughly corresponding to the Olduvai Normal Event, i.e. 1.8 million years ago. This led Raffi (1986) to conclude that the start of the coldwater conditions, more precisely of low winter temperatures in the Mediterranean occurred later than the isotope-dated cooling event in the world ocean.

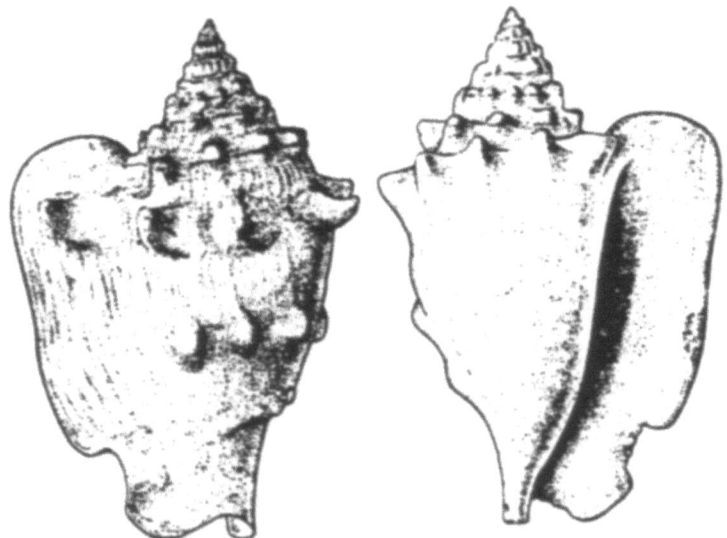

Fig. 2.3. Strombus bubonius, typical Senegalian, interglacial species in the Mediterranean.

There was a valiant effort to correlate marine glacial events with the Alpine, North European or North American glacial classification. This is the well- known system of 4 Glacials and respective Interglacials, the Günz-Mindel-Riss-Würm system with its regional variations and nomenclatures. The Mediterranean was the classical area where five elevated transgressive shorelines were described. In their sequence they were the Calabrian-Sicilian-Milazzian-Tyrrhenian-Monastirian shorelines. According to a concept best and most recently advocated by Zeuner (1959), these high shorelines were left behind by elevated sea-levels, due to eustatic rise. It was believed that during Glacial periods the global sea level was falling and conversely rising during the Interglacials. Very fittingly therefore, the four last high eustatic stages in the Mediterranean were synchronized with the three european interglacials, the Calabrian taken to represent a pre-Günz high level. Disturbing was the fact that the relative height of the four successive levels gradually decreased, as compared with the present sea level. Horowitz (1979) indicated the following levels for the Levantine shores: Sicilian +80 m; Milazzian, +60 m; Tyrrhenian +35- 40 m; the two consecutive Monastirian levels, +12–15, respectively +6 m. Similar raised beaches are known from the Lebanon.

The raised beaches of the Tyrrhenian and Milazzian contain a warm-water fossil fauna, as expected for a warm interglacial. This is the *Strombus bubonius* fauna (Fig. 2.3), named after a characteristic gastropod, today found only in the subtropical Gulf of Guinea on the west African shores. Other indicative mollusks of this so-called Senegalian fauna still live in the eastern Mediterranean: such are among others *Thais haemastoma* and *Conus testudinarius*. Initially this warmwater fauna was associated with the Tyrrhenian transgression, synchronized with the Mindel-Riss, or "great" interglacial. Modern isotopic dating, however, revealed that there are *Strombus* faunas of different ages, corresponding to different raised beach phases.

At the same time, ancient drowned shorelines were identified, corresponding to the low eustatic sea levels of the Glacials. According to Horowitz, the four low sea levels are found along the Levantine shores at the following depths: post- Sicilian, –40–50 m; post-Milazzian, –55–60 m; post-Tyrrhenian, –120–230m; post-Monastirian levels, –120–150 m.

Unfortunately, this whole system breaks down, because the continuing tectonic activity and the prevalent related uplifting of most of the Levantine shores has not been taken into account. Tectonic discrepancies made it impossible right from the beginning, to find any corresponding high shore system in the Gulf of Aqaba. In the Red Sea, too, (see Butzer, 1980) the sequence of the elevated coral reefs is very complex (Fig. 2.4).

Today the current views concerning the Quaternary Ice Ages are being completely revised. Following Shackelton and Opdyke (1976), Kukla (1980), and many more recent authors, the following picture emerges: there were some 20 periods of Glacial maxima during the last 2 million years, with an average duration of 100,000 years for each glacial-interglacial cycle. The last 10 such cycles are better known since they mostly fall within the precisely dated Brunhes Normal magnetic polarity epoch (the last 700,000 years). There is a feeling that during the last 700,000 years there was an increased general global cooling (Nilsson, 1983). Each cycle has basically the same structure: most of the cycle is occupied by the Glacial, including several warmer inter-stadials, than after a very rapid (few thousands of years!) "Termination event", the warm Interglacial lasts for 10,000 or slightly more years. (Kukla, 1980).

According to Shackelton (1975) the polar ice caps never melted completely during the Interglacials. Therefore, the sea levels during the warm periods were seldom and only slightly higher than the present sea level. The high interglacial shores would be accordingly only the result of tectonic uplifting, therefore in a

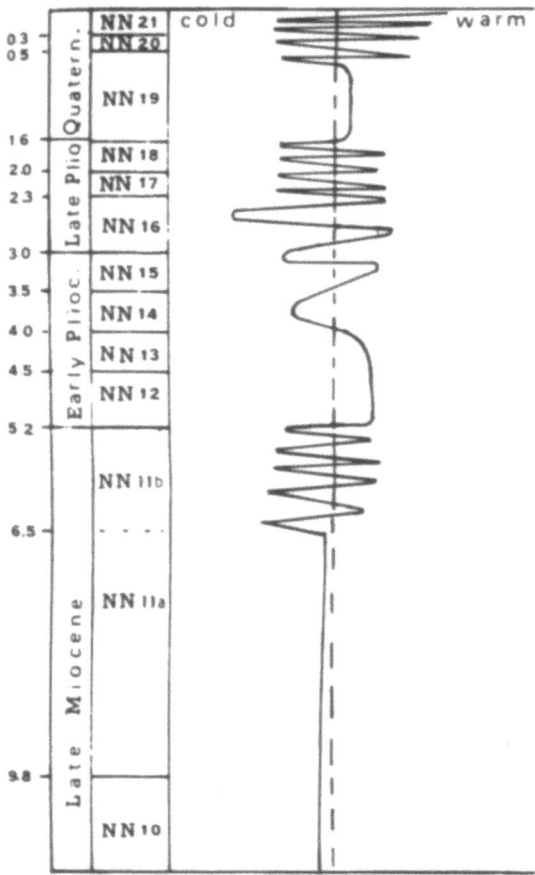

Fig. 2.4. Simplified curve of surface water temperatures in the Mediterranean from Miocene to Recent (after several authors).

tectonically active area like the Mediterranean it is only too normal that the older the interglacial shoreline, the higher it had been uplifted. For wider discussion of this subject see Hey (1978). Unlike the supposed high levels, the interglacial deep levels are real; likewise the deepening of the levels from Glacial to Glacial is a real phenomenon: it probably reflects a gradual drop of the Glacial world temperatures. According to Kukla (1981) the four classical Alpine glaciations of the many previous authors correspond in fact to four episodes of accelerated crustal movement and therefore stand out among the other glacial pulses. What conclusions can the marine biogeographer draw at this new stage of the Pleistocene research? The emerging picture is that of fairly regular, worldwide cyclic pulsations which are superimposed in each area on the local tectonic, climatologic and oceanographic circumstances. For the particular case of the Mediterranean seas around Europe, Por (1975a) developped the idea of the rhythmic alternation between two basic zoogeographical patterns, the glacial and the interglacial one. Still envisaged at that time for a "four glacial" Pleistocene, these patterns can also be used for an increased 20-fold rhythm, as seen today.

The method of taking the last glacial cycle, of the most recent 100,000 years as a model and to extrapolate it to the 2 million years of the whole Pleistocene can supply an extremely useful working hypothesis. The increasing knowledge about the last Glacial cycle acquired through the methods of isotopic and other advanced methods of dating and of paleoecologic analysis has helped to sketch-out a "prototype" of the time between two Termination events.

2.3. The model of the last Glaciation applied to Levantine marine biogeography

A first important conclusion from the presently accepted new interpretation of the high interglacial shorelines is the fact that probably there were never conditions which allowed the flooding of the Suez Isthmus. The fact that despite the maximum present elevation on the Isthmus, i.e. +23 m at El-Gisr, there were very few indications for a faunal interchange between Mediterranean and Red Sea, at a postulated +40 m level of the Tyrrhenian shoreline, was a real puzzle. Por (1978) tried to find an explanation, assuming that the Nile delta waters acted as a low-salinity barrier, an idea put forward already by several nineteenth-century authors.

Since the interglacial eustatic levels were probably never much higher than the +2–4 m level of the subrecent Flandrian transgression, representing the deglaciation climax of our present Interglacial, both the El-Gisr and the Serapeum (+10 m) ridges were not submerged, a situation which is well supported by the map of the Quaternary sediments on the Isthmus, by Fuchs (1878). The opportunities for a faunal interchange between the Red Sea and the Mediterranean were somewhat improved through the intermediary of the branches of a more active Nile and across the narrower Isthmus; but there was not a real marine contact. When the Egyptian Pharaohs built their shipping canals across the Isthmus (see Por, 1978) they utilized as best they could the opportunities of a typical Interglacial high sea-level situation.

The last Interglacial started about 127,000 ± 6,000 years ago and about 115,000–160,000 years ago the last Glacial period began (Kukla, 1980). It lasted till about 12,000–10,000 years ago, the date of the last Termination event. An international program, the CLIMAP (1976), concentrated on the study of the oceanographic conditions during the last Glacial climax, dated at 18,000 years ago.

The results concerning the eastern Mediterranean and the Levantine Basin within it are very enlightening. As shown by Thunell (1979) and by Thiede (1980), the temperature gradient between the western Mediterranean and the Levantine basin were much steeper. Today the temperature gradient is of 4.5°C in the winter (13.5 to 18.0°C) and of 5.0°C in the summer (21.5 to 26.5°C). During the last Glacial, the gradient was of 9°C in the winter (7.0–19.0°C) and of no less than

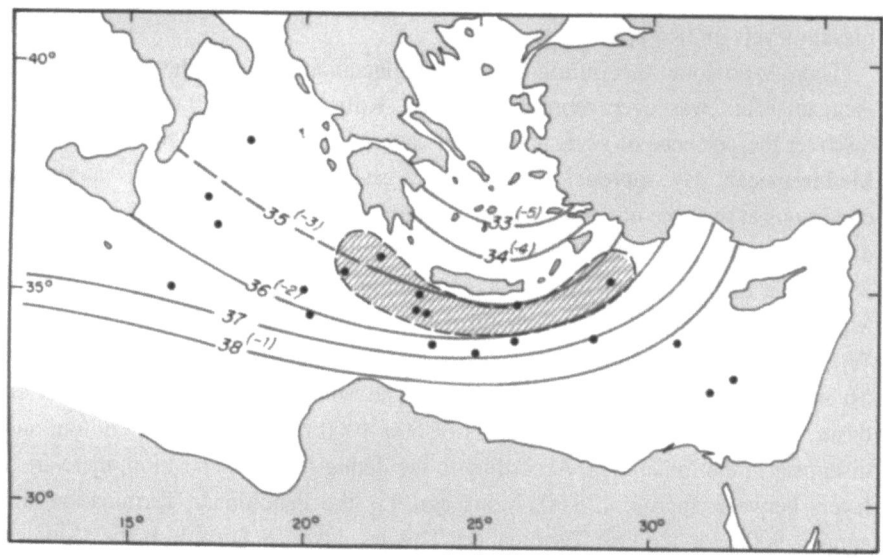

Fig. 2.5. The eastern Mediterranean during the last glacial period. Paleo-isotherms are indicated and compared to present temperatures (in parantheses). Sites of known sapropel deposits are indicated by points; shaded area covers reports of high sapropel deposition (after Thunell, 1979 and Muerdter et al., 1984).

13°C in the summer (13.0–26.0°C). In other words, the Levantine basin was only about one degree colder in the winter and perhaps only half a degree colder in the summer, compared to the present. Whereas the western Mediterranean fluctuated considerably, the Levantine basin remained almost stable.

On a closer look, however, we shall see that the Levantine basin was hydrographically separated from the Ionian basin by a tongue of cold water penetrating from the direction of the Aegean Sea. This penetration appears very clearly on the paleotemperature maps by Thiede (1980): it reached from the north to the Libyan shore of Africa. At present, there is a steep northward decrease in winter temperatures, from 17°C at the limit of the Levant Basin around Rhodes, to 11°C near the northern coast of the Aegean Sea. In the summer the gradient is much less in evidence, i.e., from 25°C to 22°C. However, 18,000 years ago, a cold tongue of 13°C reached the African coast during the winter, isolating the Levantine basin, where temperature did not reach much below the present minimum of 18°C. In Thunell's analysis (1979), this Aegean penetration is less dramatic but still remarkable enough (Fig. 2.5). Gat and Magaritz (1980) considered that such an anomaly over so small a sea-surface can only be explained through high flow rates of cold and eventually fresh waters into the area. Indeed, the influx of cold waters could only be caused by an increased flow of melt ice water from the Black Sea. Thunell (1979) indicated that this cold water was also characteristically diluted: salinities in the Aegean Sea 18,000 years ago were below 33 ppt, a decrease of over 5 ppt. The Aegean water tongue had still 35–36 ppt around Crete. At the same time there was no decrease of salinity along the Levantine shores: it stood at near-present levels of 38.5 ppt.

There is no doubt that during the rapid deglaciation events of the Terminals, the Aegean influx was even more exacerbated. Kullenberg (1965) was the first to discover the presence of several layers of sapropel in cores taken from the eastern Mediterranean. By sapropel one means a sediment which, due to anaerobic conditions at the time of its deposition, contains unusually high percentages of non-oxidized, organic carbon: the generally accepted lower limit of this concentration is 2% of the total sediment weight. A great number of authors studied the eastern Mediterranean sapropels and we are utilizing for our discussion here the following: Williams et al. (1978), Luz (1979), Rossignol-Strick et al. (1982), Calvert (1983), Muerdter et al. (1984) and Stanley (1985). The sapropels are found in the Levant basin, chiefly south of Crete at depths of 600–1000 m. They tend to thin out and disappear in the Ionian Sea. According to the dating of their deposition, there are 5 layers between approx. 125,000 years ago, i.e. the Penultimate Termination and approx. 9000, i.e. the Last Termination. This last layer (S 1) could be best studied and eventually serves as a model for the other layers. Layer S 1 and the earliest layer, S 5 are the thickest layers. The area in which maximal thicknesses of sapropel were found in the S 5 stratum, i.e. over 10 cm, are situated in an arc situated south of Crete and actually delimitates the Levant Basin towards the Aegean Sea. The sapropels do not as a rule show signs of bioturbation, meaning that they were not or scarcely inhabited by benthic fauna, evidently because of the anaerobic conditions during their deposition. Menzies (1972) mentioned that remains of several animal phyla and classes are lacking in the sapropelic deposits, such as Bivalvia, Ostracoda, Ophiuroidea, and Holothuroidea. Faunal diversity in the overlying oxygenated sediments is always considerably higher. According to Thunell (1979) the characteristic planktonic foraminiferan found associated with the sapropels is *Neogloboquadrina dutretei* a low salinity species.

The discussion around the genesis of the sapropel layers in the Levantine Basin is still undecided. Evidently, a saline stratification, like the one known in the Black Sea must have been prevalent: i.e. a low salinity surface layer obstructing vertical circulation and oxygenation of the more saline deeper layer. In addition, a source for an increased organic carbon supply must have been present.

Algal, and especially Dinoflagellate blooms, are being considered, together with an increased supply of land plant debris. Older authors saw the culprit in the restricted water exchange with the western Mediterranean, owing to a lowered Glacial sea level. Today it became clear that the sapropels were not associated with Glacial maxima: there is no sapropelic layer for the Last Glacial maximum around 20,000 B.P. One rather considers the sapropelic events to be connected with the rapid deglaciation or flooding rivers marking the end of a Glacial cycle. While everything points in the direction of a massive cold – and freshwater influx from the Black Sea, through the Aegean Sea as being the main causative factor, the influence of an increased Nile flow or at least its eventual concomitant influence must also be taken into account (see below).

Anyhow, it is now clear that for the last time around 9000 B.P and for a duration of a few thousand years (Ryan, 1972) the Levantine Basin had anaerobic conditions on the bottom caused by the combined action of an "Euxinic" type of saline stratification and a radically higher primary production. Shaw and Evans (1984) consider that along the Anatolian coast the anoxic conditions could have been prevailing already at the shallow depth of 350 m.

These conditions were much different from the conditions we know today in the Levantine Basin, which at present has one of the lowest primary productivity values measured in oceanic waters. Consequently there is much dissolved oxygen in the deepest waters, despite the high temperature of 13.7°C prevailing there.

Bacescu (1985) considers that the bottoms of today's eastern Mediterranean are often azoic and that even the bathypelagic fauna is poor and he is based on Menzies (1972) and George and Menzies (1968). These authors considered that the deep muds of the eastern Mediterranean are still inhospitable to bathyal and abyssal Isopod crustaceans. There is no doubt that the frequently recurrent anoxic events and especially the one distant only several thousands of years ago, are an important factor in the extremely poor benthic animal populations at medium and great depths. For stenohaline sessile animals such as the deep water Gorgonaria and Scleractinia the time needed for the repopulation of the Levantine Basin after a brackish and anoxic event should be kept in mind.

That Pleistocenic events could have had far-going consequences on disappearance of the deep corals in the Mediterranean is shown by Fredj and Laubier (1985), who mention that deep-sea corals disappeared from the western Mediterranean during the height of the last Glacial, 31,000 to 28,000 years ago. According to Zibrowius (1980), *Madrepora oculata* and *Lophelia pertusa* are presently in an extreme recession and most of the colonies are dead. This is also the case with the sublittoral scleractinian "reefs" of *Cladocora caespitosa* in Haifa Bay which are today composed almost exclusively of dead colonies and probably still exhibit the impact of of some fairly recent hydrographic vagaries.

The Red Sea did not suffer the influx of low salinity glacial meltwaters, unlike the neighbouring Mediterranean. Neither were there any large-scale rivers discharging their fresh and nutrient-loaded waters into this sea. As mentioned earlier by Por (1978), the once prevalent view that at low eustatic sea levels the Red Sea turned repeatedly into a hypersaline lagoon system, has been gradually discarded. In a recent publication, Klausewitz (1983) accepts the survival "in situ" of a Pleniglacial marine fauna as a necessary prerequisite for the presence of the high percentage of endemics among the fishes of the Red Sea. According to him, the Straits of Perim, which connects the Red Sea with the Gulf of Aden, was during the Glacial at least 40 m deep and 5–8 km broad and still able to prevent hypersaline extremes in the Red Sea. Even if this is not evident enough, it is clear that such a narrow Bosphorus-like connection was able to maintain repeated advance and retreat movements of fauna from and to the deep and open Gulf of Aden. Both Por (1978) and Klausewitz (1983) emphasized the fact that the Red Sea fauna is to a

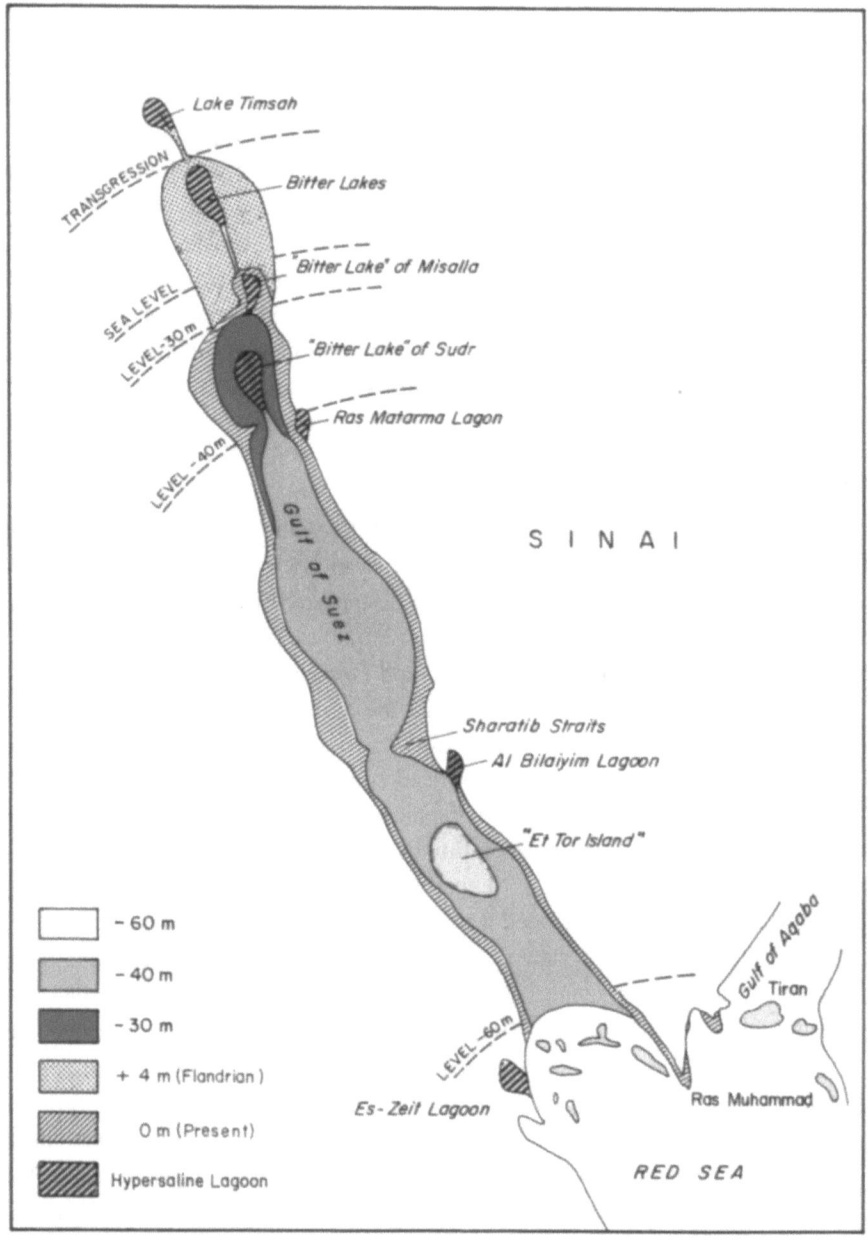

Fig. 2.6. Pleistocenic water levels in the Gulf of Suez and supposed "Bitter Lakes", metahaline terminal lagoons. Eventual tectonics are not considered (original).

considerable degree identical with that of the Gulf of Aden and even that of the Arabian Coasts (see below).

Thanks to the research projects dealing with the hot and saline brines on the bottom of the Red Sea, we know that the temperature during the last Interglacial was similar to the present one and that it declined by some 7–8°C during the last Pleniglacial (Berggren, 1969; Deuser and Degens, 1969). During the paroxism of low temperature the planktonic foraminiferans disappeared, but coccolithophorids and also one species of Pteropoda survived.

Rosenberg-Herman (1965) and recently Reiss and Hottinger (1984) focused especially on the Pleistocene conditions in the northern Red Sea and the Gulf of Aqaba. Reiss and Hottinger calculated the probable paleotemperatures and salinities for the last 150,000 years and reached the conclusion that the maximal temperature drop was not more than 4°C, i.e. to about 17°C, about 18,000 years ago. Among other evidence they consider the fact that the tropical- subtropical and symbiont-bearing foraminiferan *Amphisorus hemprichi* survived the climax of the last Glacial. This surface water species is known to be unable to survive temperatures below 15°C (Reiss and Hottinger, 1984). *Amphisorus* is known to be living as an epiphyte of the sea grass *Halophila stipulacea*. This sea grass is presently growing in the lagoons of the Persian Gulf at salinities of around 50 ppt. (Evans et al. 1973) and in the Ghor Blayim lagoon of the Sinai Coast, at around 48 ppt (Por 1978). One can therefore assume that salinity during the Glacial maximum would have been around 50 ppt. As stated by Por (1978) the Glacial Red Sea and Gulf of Aqaba were "metahaline" waterbodies (see below).

The fairly diverse fauna of the metahaline lagoon Ghor Blayim (see above) which harbours several algae, mollusks, echinoderms, marine fish and even a monospecific coral reef of *Stylophora*, all this at a summer salinity of above 48 ppt may serve as a good model for the Glacial Red Sea. Evidently, the shallow Gulf of Suez was dry at low eustatic Glacial sea levels (Por, 1978) similar to the Persian Gulf: Stoffers and Ross (1974) report Glacial levels of –105 to –125 meters in the later area. The different intermediate phases of transgression and regression probably led to the formation of different ephemeral highly saline "Bitter Lakes" in the Gulf of Suez (Fig. 2.6).

The resettlement of these gulfs after the Terminations or even during warmer Interstadials must have started with the metahaline species and then proceeded with the influx of the normal-salinity marine fauna which has been waiting "before the gates".

Reiss and Hottinger (1984) also advanced the conclusion that during the time around 18,000 B.P., there was an increase in the productivity of the Gulf of Aqaba. This might be indicated by the predominance of several species of more eutrophic coccolithophorids, foraminiferans and pteropods. The mechanism responsible for such increased productivity is not definitively clarified: the authors speak of imput from a more eutrophic World Ocean, active upwelling, longer residence time of the water in the Gulf of Aqaba and even inflow of nutrient-rich water from the continent. This latter factor could have been of importance not during the Glacial, but during the rapid deglaciation after 10,000 B.P. Interestingly Reiss and Hottinger (1984) themselves set the establishment of the extremely oligotrophic "blue sea desert" conditions in the Gulf of Aqaba as late as 4,000 B.P. This is a time which post-dates the end of the "Flandrian", postglacial temperature- and pluviosity climax and corresponds with the start of the present arid conditions.

Detailed and interesting data about the subrecent evolution of the Mediterranean coast of Israel and Sinai are contained in Neev et al. (1987). Accordingly, the period between 40,000 and 10,000 years B.P. was quiet tectonically; the sea level was lower by up to 100 m and the near-shelf and present coastal plain were occupied by fresh and brackish water lakes. Tectonic subsidence in the Levant Basin however, did not show any abatement. A short period of activ tectonism followed between 10,000 and 4,000 B.P. The Climatic Optimum in our area was according to these authors in the 6th millenium before our period. A short tectonic cataclysm in the fourteenth century (the Late Mameluk Period) gave to the coast of northern Sinai its desertic and hypersaline environments.

2.4. Pluvials, Interpluvials and lake levels

The Pleistocenic climatic fluctuations in the warm, arid and tropical zones were expressed by an alternance of wet and dry phases. The wet phases, the "Pluvials" were seen as corresponding with the Glacials and the "Interpluvials" with the Interglacials. In the most extreme form, in East Africa, four Pluvials were defined and explicitly correlated with the four Alpine Glacials.

As shown by Butzer (1978), among others, this does not reflect reality in the tropics: In North Africa and probably also in Australia and even in the tropical rainforest areas, the Glacials of the high latitudes corresponded to dry and often cooler climate. In the Sahara and the Sahel as well as in India, the last Glacial maximal cold spell of around 25,000 B.P. started cool and hyperarid conditions; these lasted at least till 13,000 B.P.

The Nile, well investigated by a series of authors, lost during the last Glacial phase its contact with the Central African lakes and the Ethiopian water supply and turned into an enormous-sized wadi, a seasonally dry river. Around 12,500 to 11,500 B.P., as rainfall in tropical Africa increased, Lakes Victoria and Mobutu (Albert) overflew and caused an almost catastrophic flooding in the bed of the dry White Nile (Williams and Adamson, 1980; Adamson et al. 1980). At some time during this Holocenic pluvial phase, even Lake Turkana (Rudolph) which has no contact with the Nile today, flowed into this river. According to Said (1981) the Nile "which entered Egypt 'hesitantly' at the beginning of the Würm glacial" turned into a gigantic river. Adamson et al. (1980) consider that the drastically increased Nile flow of the Glacial Termination was responsible for the formation of the last layer of sapropels in the eastern Mediterranean, a view already discussed above.

Since the authors tend to consider that the flow of the Fini-Glacial Nile was one order of magnitude bigger than that of today, its oceanographic impact on the Levantine Basin should be briefly mentioned, based on an actualistic approach.

The Nile flood carried by the Coriolis force and the longshore current, flowed east- and northward along the Levantine coast. In the 1930's when the hydrographic measurements started in our area, the Nile was already closed behind the first Aswan dam and there were gates on the two Nile Delta branches, the Rosetta and Damietta branches. These gates were usually opened in August at the time of the seasonal Nile floods. According to Liebman (1935) the flood current took 6 weeks to reach the Israeli coast. In 1930, the salinity was lowered to 34.02 ppt as far as Haifa (Oren, 1969) i.e. more than 5 ppt lower than normal Levantine surface salinity. In 1947 at Gaza the Nile water diluted the inshore waters to 25 ppt, i.e. a difference of over 14 ppt from normal (Oren, 1969). Decreased salinity values were recorded in the months of September and October as far north as Beyrouth.

We have every reason to assume that at a tenfold increased Nile flow, the diluting impact was much stronger, more long-lasting and far-reaching: At that time the White Nile (i.e. Central African lakes) component was proportionally larger than the highly seasonal Blue Nile component. As a consequence, the increased outflow must have been also much less seasonal.

Adamson et al. (1980) also mention that in the early Postglacial times the buffering of the White Nile by the Sudd swamps had not yet fully developed. All in all, we imagine that the salinity fluctuations in the surface waters along the Levantine coast were very much larger than in recorded times. Whether the Nile water dilution was enough to account for the stratification needed for the sedimentation of the sapropels is doubtful (see above). However, it was strong enough in order to be an important factor in inducing faunal impoverishment along the Levantine coast.

The sharp increase in pluviosity at the last Termination event, more precisely around 12,500 B.P. is known to have also caused a steep increase in the water levels of the Subtropical African Lakes. The case of the famous "Mega-Chad" lake which turned into an enormous inland sea during the early Holocene, is classical (see for instance Servant and Servant-Vildary, 1980). Also a whole series of Ethiopian lakes, including the lakes of the desertic Afar triangle, exhibited very high water levels immediately after the start of the deglaciation (Gasse and Delibrias, 1977). Some lakes even formed in the Rub el Khali the "Empty Quarter" of the Saudi Arabian desert (McClure, 1976).

In general over the whole globe and in particular in the African deserts there followed a sharp downward trend in temperature and humidity, around 5,000 B.P. After that time, present conditions became more or less established. The several thousands of years of increased fluvial activity and high lacustrine levels must have played a crucial role in the distribution of the freshwater biota in desert areas. This has been very convincingly analysed for the Saharian freshwaters by Dumont (1979). Without doubt, the Pluvial event around 12,000 was a relatively short-lived happening. So most probably were all the other, previous Pluvials, if, as it seems, they were always associated with the short and steep deglaciations following the Terminations. These were pulsations accompanied by repeated colonizations or retreats but their duration was probably not enough to allow for speciation in the newly settled environments, before the next drying-out. Dumont (op. cit.), emphasizing this important aspect, believes that some 20,000 years of uninterrupted colonization of an isolated desert oasis would be necessary for a morphologically visible speciation step.

Fairbridge (1972) considered that the above-presented "African" climatic phases are also to be found in the late Pleistocenic-Holocenic history of the Levantine lakes. In contrast, Israeli authors like Neev and Emery (1967), Horowitz (1973), Begin et al. (1974), Issar and Bruins (1983) as well as expeditionary studies (Kaiser et al. 1973) strongly advocate the existence in the Levant of "northern Pluvials" i.e. increased humidity during Glacials. Farrand (1971), based on a comparison of the lake levels in the Middle East reaches the conclusion that there was a wet-dry-wet sequence during the last 50,000 years of history in the Levant.

The level of Lake Konya in Anatolia was highest around ± 20,000 B.P. (Roberts et al. 1979). It now seems that while northern Africa and even Arabia were going through a very arid phase around 18,000 B.P., lacustrine basins in the Levant and Asia Minor exhibited on the contrary high water levels. Attempts to apply the very well-documented and modern paleolimnological conclusions obtained from African studies to the Levant situation (see for instance Gat and Magaritz, 1980) are not very convincing.

The Levant was climatically "out of phase" with the Saharian climatic events. Rognon (1979), based on an actualistic example supplied by Rosenan (1951), considers that when the Saharian anticyclon moves to the south, thus preventing the wet monsoons from reaching the Sahara, improved conditions for winter rains develop in Israel. It follows that during the cold-dry Pleniglacial conditions of the Sahara, cold-wet conditions might have prevailed in our area.

The factual information we possess is based on studies of the Dead Sea (Neev and Emery, 1967; Begin et al., 1974); swamps and spring travertines in Sinai and the Negev (Issar and Eckstein, 1969; Issar and Bruins, 1983), of Lake Hula (Picard, 1963; Cowgill, 1969; Racek, 1974 ; Horowitz, 1978; Ohlhorst et al. 1982), of the Pleistocenic Damascus Basin lake (Kaiser, et al. 1973) (Fig. 2.7).

The present desertic conditions in southern Israel and in Sinai would have started only around the 7th century of the Modern Era (Issar and Tsoar, 1987; Issar et al. 1987).

Horowitz (1988) gives an up-to-date summing-up of the sedimentary history of the Jordan Rift Valley. Accordingly, up to the end of what he calls the "Preglacial

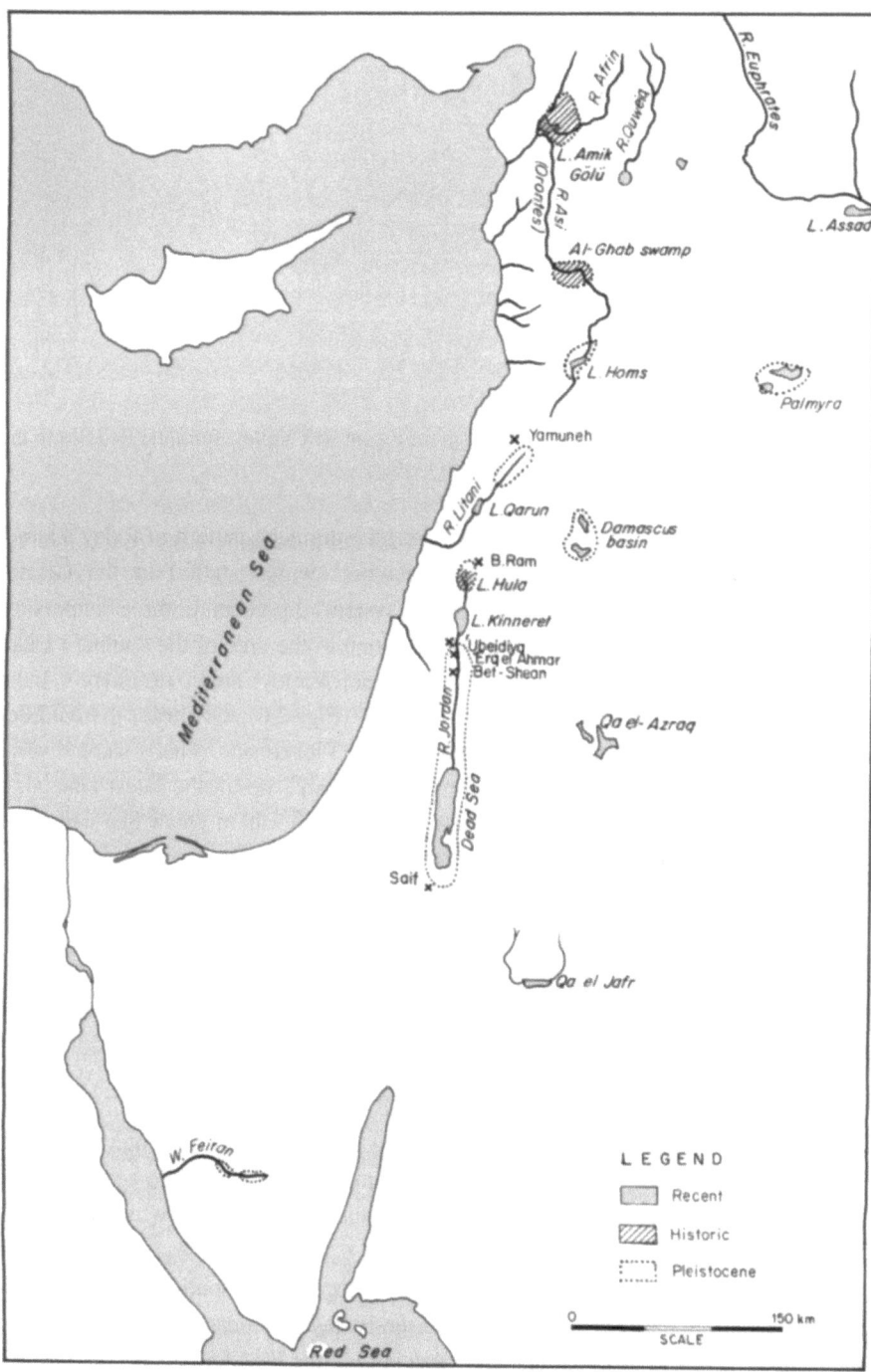

Fig. 2.7. Bodies of standing water in the Levant, during the Pleistocene, in historical times and at present (original).

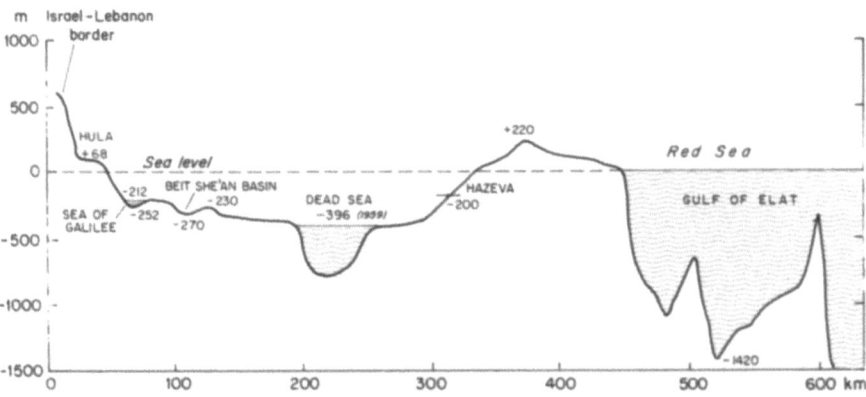

Fig. 2.8. Topographic cross section through the recent Rift Valley, from the Red Sea to the Lebanese border (after Neev and Emery, 1967).

Pleistocene" (i.e. about 1.7 million years ago) the drainage pattern of today's Israel was different (see above). The rifting tectonics which started in the Glacial Pleistocene, formed the endorheic Jordan system. In general, three centers of tectonic subsidence were active in the valley: one in the area of the southern Dead Sea and south of it; one in the Central Jordan Valley, south of today's Lake Kinneret; and one corresponding to Lake Hula (Fig. 2.8). Of these, the Middle-Jordan center remained inactive for most of the Pleistocene. There were several subsidences in the other sites, but during the "Würm", i.e. in the latest Glacial, a very marked subsidence created the extremely deep northern Dead Sea and Lake Kinneret. In conjunction with the tectonics, there were frequent extrusions of basaltic magma, over large areas adjoining the Rift Valley and sometimes daming it.

Because of the repeated and non-synchronic tectonics and the basaltic outflows, the picture of the lacustrine levels as an expression of a humid-arid alternance is considerably complicated. Lake Kinneret, for instance, is unanimously considered to be a very new lake, owing its existence to the latest Würmian or even post-Würmian subsidence in the Jordan graben. Basalt outflows probably had a considerable role in daming lake Hula, reducing or even at times interrupting its drainage to the Middle Jordan. The basin of Damascus with its shallow Pleistocenic lakes came into being only after the outflow of the Hauran basalts which isolated it from the Jordan graben rivers.

According to Bender (1968) there was a Pleistocene lake south of the Dead Sea, at Wadi Fidan. In eastern Jordan in the presently desertic areas, there were lakes in the tectonic valley of Wadi Sirhan and especially in the El Azraq area.

In the Lebanese-Syrian sector of the Rift Valley, the basaltic outflows, combined and connected with the tectonics had a similar, though less well-known role. The main reason for the lack of detail in the knowledge, is the fact that for most of the events there is no precise radiometric dating.

The encyclopedic work of Dubertret and his co- workers was the basis for several more specific and recent publications on the Syrian-Lebanese area, which we used. Such are De Vaumas, (1957a, 1957b), Van Liere (1960/61) and Wolfart (1967). Following these authors, there are several crucial areas of basaltic extrusions across the Rift Valley. The northernmost is responsible for the daming of Lake Amiq (Lake of Antiochia) whereas more to the south, large basaltic areas between Aleppo and Homs, isolated the Proto-Orontes from the now endorheic basins of Queiq and El Bab. North of Homs, near Rastan, another basaltic dam crosses the Rift Valley. The age of the youngest basaltic flows seems to be even more recent than in the Jordan Rift: 4000 B.P. (De Vries and Berendsen, 1954).

Because of these basaltic barriers the Syrian Rift Valley at times harbored three temporarily endorheic basins: that of Lake Amiq, before the breakthrough of the Orontes outlet to the sea; the Ghab valley, north of the Rastan sill and the basin south of it, corresponding to the basin which contains the lake of Homs. Another lacustrine basin existed in the Lebanese Bekaa, especially before the opening of river Litani's drainage to the Mediterranean.

Because of the above-mentioned new methods of Pleistocene chronology, most of the conclusions which have been correlated with the "Four Glacials" system of Europe will have to be revised. The sequence of events which occurred in the Levantine aquatic network can be retraced with some confidence only for the last Glacial cycle: The Last Interglacial was represented in the Dead Sea area by a shallow lake (or lakes) usually named the "Samra lake" (Begin et al. 1974). During the last Glacial a very large lake expanded in the Levantine Rift valley: Lake Lisan. Situated at about −180 m below sea level and with depths of at least 190 m, Lake Lisan was in existence between 60,000 and +/−18,000 B.P. (Neev and Emery, 1961). Several authors like Kaufman (1971) divided the period and spoke of a more shallow Lower Lisan Lake between 60,000–40,000 and a subsequent deep Upper Lisan Lake.

At its height, this Late Glacial lake was 220 km long and probably less than 17 km broad, a replica of some East African rift lakes. It reached some 20 km south of the present Dead Sea and in the north reached the southern corner of the present Lake Kinneret. According to Begin et al. (op. cit.), the lake was incompletely divided by a sill, south of the present Bet Shean valley. Lake Lisan was considerably less saline than the more than 300 ppt brine of the recent Dead Sea. The salinity values were inferred from the Diatomacean fossils analysed (Begin et al. op. cit.) and also from the few other fossil remnants, among them some fish. The Lake had a marked positive N-S salinity gradient with maxima of around 150 ppt. Lake Lisan ended fairly abruptly around 18,000 B.P. and several authors connect this with the last tectonic activity in the Rift valley, which lowered the bottom of the graben by some 300–400 m, to the present −794 m. Based on volume calculations, Neev and Emery (1967) reach the conclusion that there was a loss of 136 cubic kilometers of water, from the original 325 cubic kilometers of Lake Lisan: therefore tectonic deepening alone cannot explain the lowering of the lake level and its reflux into the deep Dead Sea basin: net loss of water must be taken into account, too. This can, of course, also be corroborated from the probable salinity differences between the old lake and the Dead Sea. That the shrinking of Lake Lisan was at least in part climate-controlled results from the fact that around 10,.000 B.P., i.e. more or less during the Termination event, there was again a short but marked increase in the level of the Dead Sea (Neev and Emery, 1967). Actualistic considerations (see above) also point in this direction.

With the Dead Sea receding into its present bed, the Jordan started to flow and erode the Lisan sediments. For some time a lake existed in the Bet-Shean area, but was drained by the down-cutting Jordan around 5,000 B.P.

Lake Kinneret is not much older than the Glacial Termination. South of the Kinneret, older shallow lake environments are known: The Erq-el Ahmar lake which closely overlays the Cover Basalt event (see above) and is therefore considered to be somewhat younger than 3 million years. The Ubeidiya Lake probably existed for a considerable period of time, before the Brunhes normal magnetic epoch, i.e. earlier than 700,000 B.P. For these early lacustrine phases, see among others Tchernov (1987).

Lake Hula (also spelled Lake Huleh) was drained in 1958 (Dimentman et al., in press). In 1963 several cores were taken in the old lake-bed by Cowgill (1969): one of the cores, 54 m long, made possible the study of the last 30,000 years of the Lake. While it was known for long that during the Middle-Upper Pleistocene the area was occupied by several successive swampy lakes, the study of the Cowgill cores (Cowgill, op. cit.; Hutchinson and Cowgill, 1973; Racek, 1974; Olhorst and Hutchinson, 1977; Olhorst et al., 1982) showed that around 30–20,000 B.P. the shallow swampy lake turned into an extensive and fairly deep lake. A maximum depth of 10 m was eventually reached around 16,000 B.P. This event was probably not caused only by the daming of the Hula basin by basaltic extrusions, as believed by Picard (1963) (his Upper Lacustrine Unit). Wet-cool glacial conditions also acting upon the contemporary Lake Lisan must have been the main reason for the high water level of Lake Hula. The basaltic obstruction was much older. Cowgill identifies the basaltic dam with the Yarda outflow of the literature; but this has been recently dated between 560,000–850,000 B.P. (Tchernov, 1987). The latest basaltic layers in the area are the Hasbani basalts which according to Horowitz (1978) are dated 70,000–80,000 B.P.

From 16,000 onwards, the lake gradually shallowed and changed its regime back to that of a swampy lake. In this process the water chemistry modified from slightly alkaline to low and variable pH regimes. This evolution is fairly well documented by the fossil sponges of the lake (Racek, 1974) the mollusks (Olhorst and Hutchinson, 1977) and fishes (Ohlhorst et al., 1982).

According to Horowitz (1978) the conditions in which the Early Postglacial Ashmura formation was laid down, indicate again a more "pluvial" climate. Present-day drier climates and a final shrinking of lake Hula seems to have started around 4,500 B.P. Horowitz's climatic scheme is based on pollen analysis and therefore some care should be taken when it comes to extrapolate it on the fate of relatively small lake areas which do not always faithfully reflect the changes of the terrestrial vegetation of much larger terrestrial areas. The Pleistocene lake which filled the Qa El Jafr depression reached according to Farrand (1971) a size of 1000–1800 square kilometers (comparable to Lake Lisan). It dried out around 26,000 B.P., then passed again through a swampy freshwater phase in the latest stages of the Würm before turning into the recent playa.

In the endorheic Damascus basin, river Barada and Nahr el Aouaj run out in terminal salt swamps, Bahr el Ateibe, respectively Bahret el Hijjan. According to Kaiser and his coauthors (1973) the basin was filled at least twice, by a large and shallow lake: in the Middle-Pleistocene and more recently around 20,000 B.P. The Würmian lake indicates in the view of these authors the existence of a cold but wetter climate during the last stadial of the last Glacial. During the warmer interstadials the lake shrunk. Its fate was similar after the end of the Glacial. There was a short period of higher humidity, expressed in the sediments of the Damascus basin, around 4000–5000 B.P. There is good evidence for this "out of phase" rhythm, from the palynological data (Leroi-Gourhan, 1973), mollusk fauna (Schütt, 1973) and ostracods (Kempf, 1973). The shallow Damascus lake at the time of its maximum extension covered an area of about 2,200 square kilometers.

Thus it appears that during the period ranging roughly between 50,000–20,000 B.P., the Levantine area was covered by a series of large-sized lakes; they shrunk considerably or dried out near the Termination and experienced a brief and not very impressive recovery during the Postglacial "climatic optimum" before disappearing or shrinking to their present levels and sizes (Fig. 2.7 above).

2.5. Shifting rivers and captured headwaters

The combination of tectonic activity and climatic pulsations must have had a considerable influence on the fluvial catchment basins of the Levant. We have already mentioned that the Jordan became endorheic largely because of the Pleistocene uplifting of the Cis-rift mountains. Its different sectors underwent changing fates and at least once during this period, the Jordan was submerged for most of its course by the waters of the Lisan lake.

There were probably repeated switchboard contacts between the east-west rivers flowing into the Mediterranean and the north-south rivers of the Rift Valley. Much information about these changing riverine contacts is found in Kinzelbach (1980, 1987), Kinzelbach and Roth (1984) and Krupp (1987).

As shown by these authors, the older Oriental contacts to the Euphrates system were gradually lost (see below). But their initially prevalent role explains the fact that the Levantine fish fauna is overwhelmingly Oriental in its origin (Banarescu, 1970).

For some as yet undetermined time, the river Litani of today flowed into the

Jordan system, as a mere tributary, like river Hasbani. At some stage through headwater capture or more probably through tectonic movements, the Litani changed its course and turned sharply to the west, flowing into the Mediterranean. In the area of the Bekaa valley there is today a considerable gap between the headwaters of the Litani and those of the Orontes. Shortlived aquatic connection must have existed here or through the intermediary of the Damascus basin, to allow for faunal exchange and thereafter produce populations of mollusks and decapods differentiated to the subspecific level (see below).

River Orontes was originally and/or temporarily separated into three unconnected rivers: The upstream segment, from the headwaters to the basalt dam near Homs; the middle sector corresponding to the present Ghab and the lower course which today opens to the Mediterranean. The lower course is probably of later Quaternary origin (Van Liere, 1960/61) and it resulted from tectonic subsidence and the capture of the River Kutchuk Assi (or Ak Su), the river which drained lake Amiq.

A contact between the upper segment of the Orontes and the Mediterranean coastal river Nahr el Kebir South must have existed some time ago: this seems to be evident on biogeographic grounds. The contact between the early Orontes and the coastal river was interrupted by the late Pleistocenic basaltic flows of Homs. At some stage the Orontes was also connected to another Mediterranean tributary, the Nahr el Kebir North.

For some time river Qwaiq, near Homs, was a tributary of the Euphrates. The Qwaiq probably also had alternating contacts with the Orontes. Today the Qwaiq is reduced to an endorheic situation and runs out in the desert, forming a salty swamp.

In the present Gulf of Iskenderun the Lower Orontes could establish contact at low eustatic sea levels with the Anatolian River Ceyhan. According to Kinzelbach (1980) the present course of the Orontes, composed of its different sections, became established around 6,000 years B.P. (see below). Two other Anatolian rivers, the Afrin and the Kara-Su which flow into the now drained Lake Amiq might have flowed in some earlier times in the direction of the Euphrates too. Possible cases of headwater captures and faunal interchanges are also being postulated by Por (1975b) between the coastal rivers of Israel and the Jordan system.

As suggested by Tchernov (1988), low eustatic sea levels might have facilitated the interchange between the freshwater faunas of the Nile with Wadi El Arish in northern Sinai and eventually also with some coastal rivers of southern Israel. Biogeographical aspects resulting from these modifications in the drainages of the Levantine rivers will be discussed extensively in Chapter 6.8.

There is no doubt that combined with the changing levels of the lakes, some of them even ephemeral, the permanently modified courses of the Rift Valley rivers created a very complex network of "steeple-chase" contacts (Por, 1975b). There is little chance that more precise chronologies will ever be established for these biogeographically important river captures and these frequently shifting water divides. Only the distributional patterns of primary freshwater animals can document the existence of the different changes in the river networks and the catchment basins. Eventually the ecological requirements of the species and the degree of evolutionary differentiation between the populations can further elucidate and pinpoint the hydrographic changes. Yet, in some cases interesting details can emerge from very careful correlation between geological structures, correct dating within the now widely accepted scale of isotopic stages and the pollen data. Such is the recent discussion centering about the genesis of the travertines and lacustrine deposits of the Pleistocenic Lake Saif in Central Negev. First mentioned by Sneh (1974), these freshwater deposits surrounded today by a forbidding desert environment near the western escarpments of the Rift Valley have been variously dated for the Würm or for the Riss glaciations. It seems, however, that the most active travertine deposits are of isotopic stage 7, i.e. ± 250,000 years B.P. and stage 5, i.e. about 100,000 B.P. Livnat and Kronfeld (1985) see in them a proof for humid conditions in southern Israel during these stages. Horowitz (1987) retorts by postulating a connection between the Saif travertines and the phases of tectonic activity in the Rift valley. In his words, an interesting correlation can be seen between active faulting movements and the peaks of freshwater vegetation pollen. He illustrates this data from a borehole in nearby Amatzia. The faulting eventually exposes buried aquifers and increases spring activity. Thus, in a tectonically active region, streams and consequently lakes might reach peak phases even during a worldwide or regional arid climatic stage. As shown by the answer by Livnat and Kronfeld (1987) to Horowitz's argument, the subject is still in a phase of adjustment. As it stands, it is a very stimulating hypothesis to think that tectonics have influenced the continental water network of the Levant in still another way besides modification of the topography. It seems evident that the climatic evolution of the Pleistocene Levant was different from that of Saharian Africa and the Arabian Peninsula. The role of the "non-climatic factors" emphasized by Rognon (1980) for the Saharian lakes and drainages was much more important in the Levant. It seems that at times tectonic and eruptive events gained the overhand over the climatic changes in defining the evolution of the lakes and rivers of the Levant. Farrand's (1977) conclusion that tectonic activity in the Jordan graben might have influenced the local climate more strongly than the distant glaciations should be kept in mind despite the probable overstatement it represents.

That the discussions around the different local climatic regimes are still far from being settled is shown by other authors, who accept a scheme of late Pleistocenic climatic fluctuations based on the palynological data of Van Zeist and Woldring (1980) from the Ghab valley in Syria which contradicts the scheme presented by Horowitz (see above). Again, local tectonic evolution could be responsible for the rather detailed climatic events, especially when they are based on pollen analysis.

3. Eastern Mediterranean

3.1. The Levant Basin

Margaleff (1985) fittingly characterized the Mediterranean as being "an extremely complex residue of a very dynamic past". For such a relatively small sea, the diversity of hydrographic regimes and of topography is without equal. At least 8 basins or "seas within the sea" are in evidence, not taking into account the satellite Marmara and Black Seas. Kiortsis (1985) recognizes as many as 13 or even 14 subregions within the Mediterranean proper.

The separation into a western and eastern Mediterranean along the Sicilian-Tunisian sill does not simplify the picture. On the contrary: the western Mediterranean stands out versus the complex eastern part as a relatively homogeneous and geologically similar area. Unfortunately, too many biogeographical treatments follow this inadequate binary division.

The "eastern Mediterranean" contains the shallow Adriatic Sea with its relatively low winter temperatures and low salinity; the Sea of Sidra (also called sometimes "southern Mediterranean"), shallow with relatively high temperatures; the Ionian Sea, extremely deep; the Levantine Sea, deep with high salinities and temperatures, and the Aegean Sea which is shallow and has low salinities and temperatures and furthermore suffers the influx of the Black Sea. Future biogeographic treatments should be more explicit rather than mixing all these basins into one bag as "eastern Mediterranean".

The Levantine Sea or rather Levant Basin which is our present object can be defined in several ways. In geographical terms it represents the south eastern part of the Mediterranean, separated from the main body of this sea by a zone which connects Cape Matapan and the coast of Cyrenaica,or roughly by longitude 22° E. From the shelf sea of the Aegean, the Levant basin is separated by the island arch extending from Crete to Rhodes. In a more detailed treatment of this subject (Por, 1978) it was proposed that the annual mean surface isotherm of 20°C and the surface isohaline of 39 ppt should define the Levantine basin. In view of the frequent and very extreme historical fluctuations in the hydrography of this part of the Mediterranean, we now believe that a more "neutral" geographic definition of the Levantine basin would be appropriate.

More recently Por (1981) suggested the existence of a "Lessepsian biogeographic province" which roughly corresponds with the Levantine basin. This view emphasized the admixture of Red Sea species which immigrated through the

Suez Canal, to the autochthonous stock of the Mediterranean. It is indeed possible that, in the long run, the biogeographic limits reached by the Lessepsian migrants will better define the Levantine sub-unit of the Mediterranean. However, the settlement and adjustment of the immigrants is still going on. Also the biological changes caused by the cessation of the Nile flow are still unfolding.

Considering that the Levantine basin is the oldest part of the Mediterranean (see above) and that events of such a far-going importance are still at work in it, it is worthwhile to quote again Margaleff (1985): "Geologically, the Mediterranean area has always been in turmoil, with upheavals and collisions... So many elements of stress and friction have accumulated, that we can foresee that the Mediterranean will remain active for a long time, both geologically and humanly." If this is true for the Mediterranean as a whole, it should be considered even more emphatically for the Levantine basin.

3.2. Some oceanographic conditions

Hydrographically, the Levantine basin is best characterized by the attributes of the Levantine Intermediate Water which is formed here and flows out into the western Mediterranean. According to the most recent updating by Hopkins (1985), the LIW formed near Cyprus and south of Rhodes has a temperature of 15.7°C and a salinity of 39.1 ppt. In the water column the Levantine Intermediate Water soon after its sinking down, occupies the depth layer of 200 to 700 m (Oren, 1983).

The Levantine Basin is for most of its area characterized by depths of over 2000 m and narrow shelves. More than 4000 m are reached in the Rhodes Trench and around Crete, where the bottom falls gradually to the maximal Mediterranean depth of 5093 m in the Matapan Trench.

Deep-water temperatures of the Levantine basin are relatively higher than in the rest of the Mediterranean: 13.7°C compared with 12.8°C in the western Basin (Fredj and Laubier, 1985). As first shown by the measurements of the Austrian "Pola" Expedition, the oxygen levels in the depth of the Levantine Basin are very high, i.e.up to 75 % of saturation.

At the surface, temperatures fluctuate in a relatively narrow range: in the winter the lowest temperatures are in the range of 16°C and in the summer they rise to 29 °C (Por, 1978 fide Oren). Salinities may reach a very high value of 39.5 ppt (corresponding to that of the Levantine Intermediate Water), but in general are everywhere above the 39 ppt mark. The only exception was found offshore from the Nile delta from where during the late summer-autumn months a tongue of diluted water extended along the Levantine coast, reaching as far as Lebanon (see Fig. 3.5).

The salinity gradient is extremely steep around the longitudes 21–22° E where in the summer the surface salinity increases from 38 to 39 ppt (Fig. 3.3).

Recently data on the mesoscale have been found, relative to the oceanographic

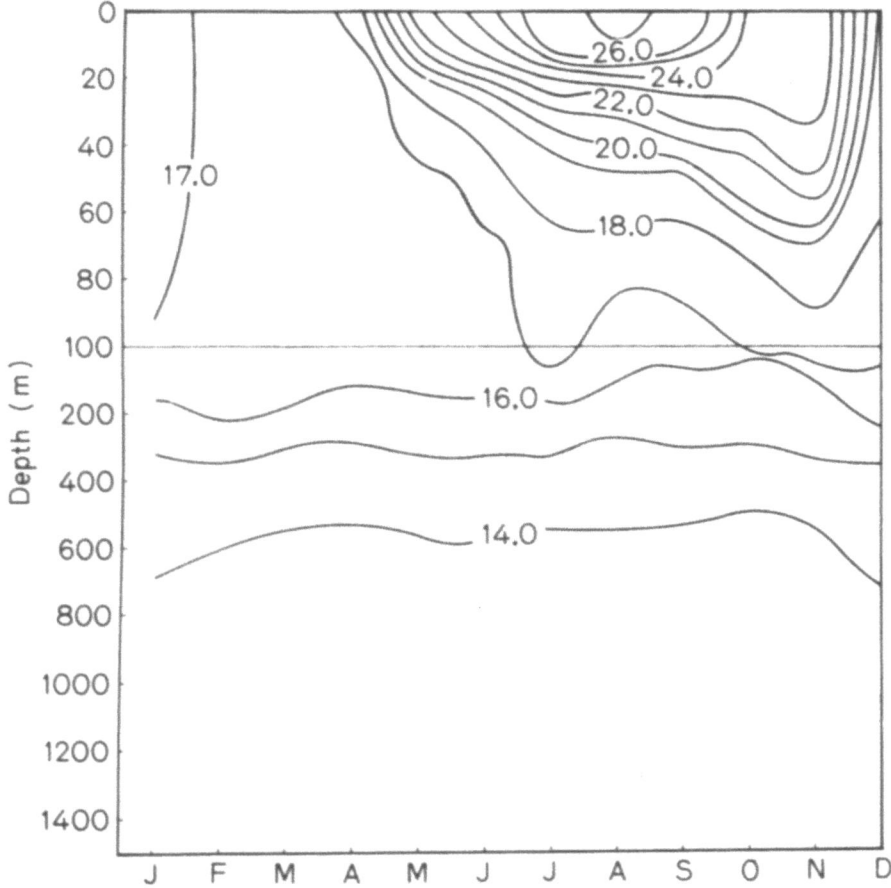

Fig. 3.1. Vertical distribution of temperatures in the Levant Basin (Oren, unpublished).

parameters in the Levant Basin. A. Hecht, S. Brenner and M. Krom (Biennial Report 1986–1987, Israel Oceanographic and Limnological Research, Haifa) report the finding of several anticyclonic warm-core eddies in the area. Water of higher temperature and higher salinity is down-welling in these areas and oxygen and nutrient curves also accompany the gyre centers down to several hundreds of meters depth. A permanent eddy is situated southwest of Cyprus and another, near the coast of Israel (Fig. 3.3). The significance of these major areas of oceanographic dynamism for the Levantine biota is not yet analyzed.

The Mediterranean as a whole has very low productivity levels. In this, again, the Levantine Basin excells. The severe oligotrophic conditions of the Levantine Basin have repeatedly been used by many authors in order to explain the impoverished species diversity of even the small sizes of the Levantine biota. No convincing proofs for a direct influence have as yet been provided. The shores are

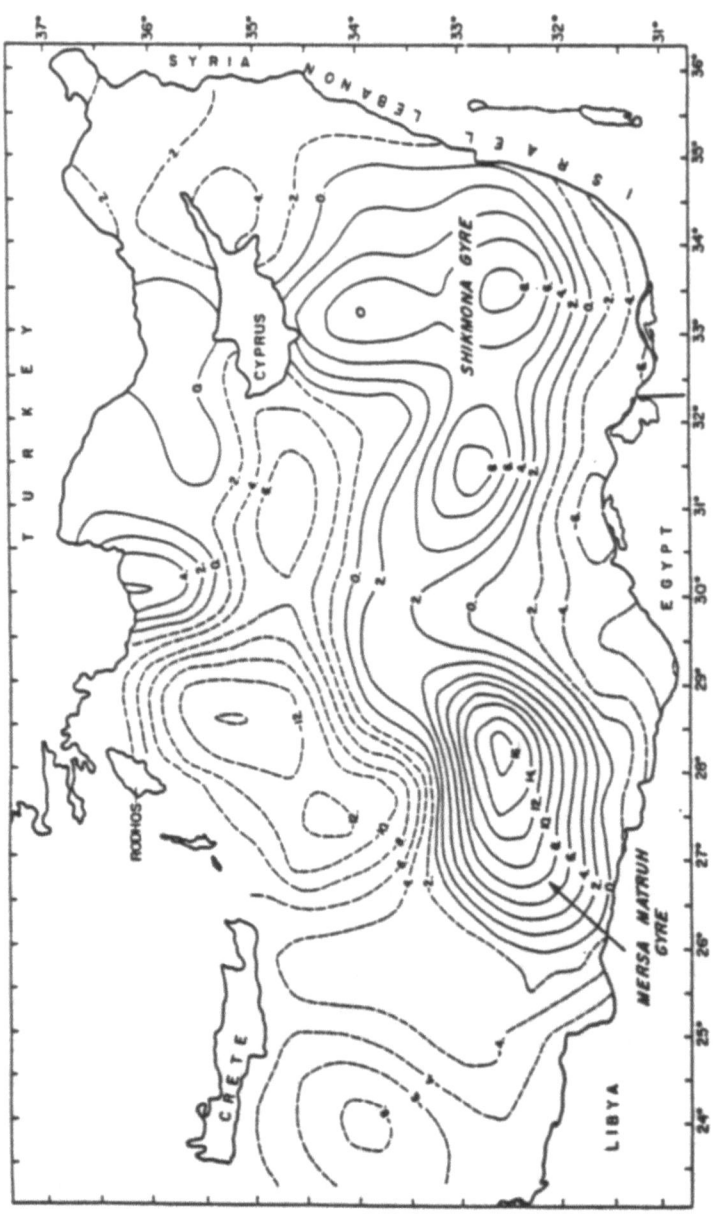

Fig. 3.2. Anticyclonic warm-core eddies in the Levant Basin, in dbars, October-November 1985 (from A. Hecht, S. Brenner and M. Krom, see p. 52).

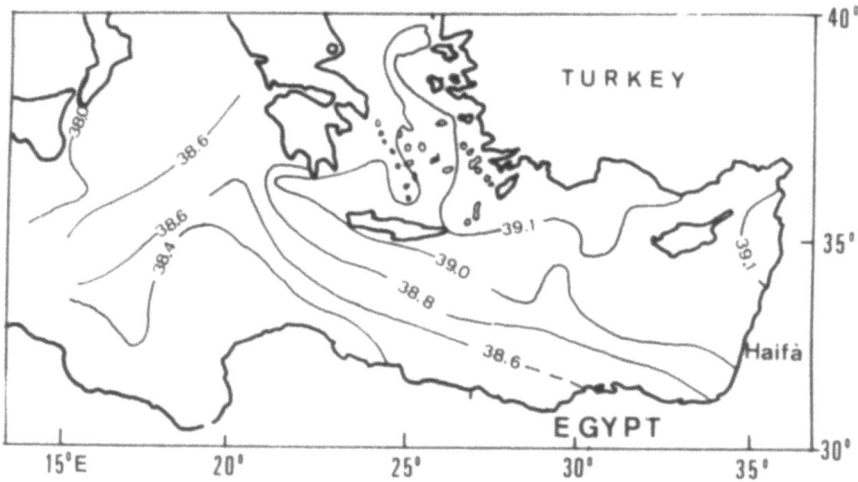

Fig. 3.3. Surface salinities in the eastern Mediterranean (after Lacombe and Tchernia, 1960).

predominantly sandy and sedimentary for the whole stretch extending along the African coast and the Levant coast up to Haifa or even the Israel-Lebanon border. This means that about 2/3 of the coasts are carpeted by (mainly Nile) sediments and present very few rocky shores.

South of Rosh Haniqra (Ras el Naqura) on the Israel-Lebanon border the shelf is built up by Nile sediments, relatively poor in carbonates. Till 1965 (prior to the Aswan dam), the Nile discharged an average annual 35×10^7 cubic meters and about $100-300 \times 10^6$ tons of sediment. The shelf is very broad in front of the Nile delta: up to 75 km (Gorgy, 1966). It gradually narrows to about 30 km off southern Israel and 10 km near Haifa. Along the Lebanese and Syrian coasts the shelf is extremely narrow to practically non-existent. Only the Gulf of Iskenderun in the northeastern corner has a relatively well-developed shelf. According to Nir (1982) the whole stretch from the Nile to Akko (Acres) is one large cell of Nile-origin sediments. Along the Lebanese and Syrian coasts and presumably on the Anatolian coast, the carbonate content of the littoral sediments gradually increases, from less than 10% along the Israel shores to an average of 31% (Nir, op. cit.).

Correspondingly, the continental slope, a very gentle 2° along the Sinai coasts becomes steeper along the Israel coast: ± 8.5° (Emery and Bentor, 1960) and is exceedingly steep along the Lebanese coast, where mountains descend abruptly into the sea and deep-sea trenches, perpendicular to the shoreline reach within a few kilometers from the coast.

According to Vanney and Gennesseaux (1985) there are about 10 such deep canyons, and in that area the shelf is only about 10 km wide. The canyons descend abruptly to 1,500 m depth, the depth of the so-called Damietta cone.

Fig. 3.4. "Kurkar" sandstone ridge along the Israeli shore (from Nir, 1982).

The Egyptian coast, Sinai included, has no rocky outcrops and there is a chain of big lagoons along the coast. In southern Israel beachrocks appear and around Tel Aviv ridges of aeolanitic sandstone ("kurkar") become prevalent. The kurkar ridges may rise to +50 m. There are several more belts of kurkar at different depths on the shelf, delimiting Pleistocenic low sea levels. A number of small islands built of kurkar accompanies the shore. According to Nir (1973) the sandstone cliffs are easily abraded and retreat landwards at a high rate of 3 cm/year (Fig. 3.4).

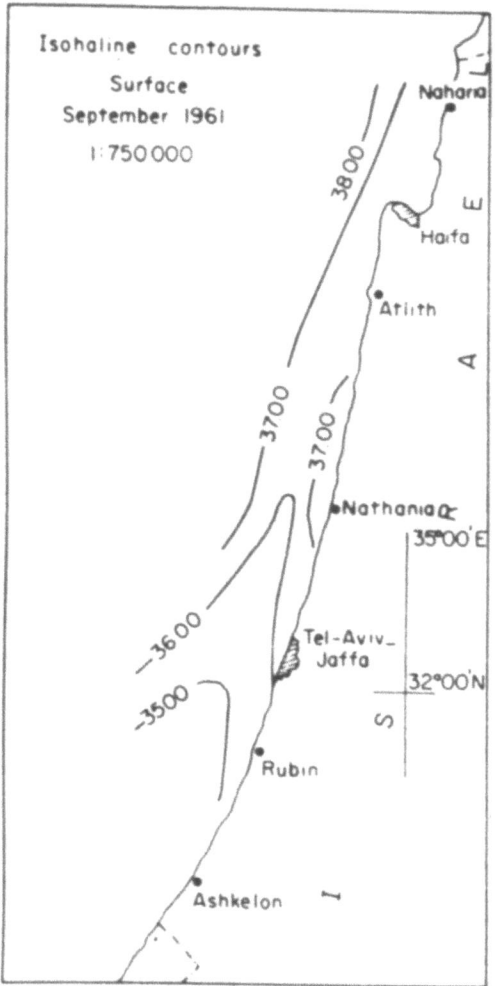

Fig. 3.5. Isohalines of the Nile flood of 1961 along the coast of Israel (after Oren, 1969).

Starting south of Haifa, the Carmel mountain ridge reaches the sea. North of Akko the shore is almost constantly accompanied by the longshore ridges of the Lebanon and Ansariye mountains. Along the Lebanese and most of the Syrian coast, calcareous rocks alternate with kurkar ridges. In northern Syria and Anatolia igneous and ophiolitic formations also reach the sea (Sanlaville, 1982). Accordingly, the sediments are also increasingly mixed with dark minerals.

The coastline is relatively straight, the only slightly larger bays of the Levantine coast are the Bay of Akko (Acres), the Bay of Akkar (Syria) and indeed the Gulf of Iskenderun. The kurkar islands along the shore are in some cases connected (also artificially) with the shore through sandy tombolos. But besides these islands, the only large island of the Levantine Basin is Cyprus.

Though not well surveyed, it seems that there are relatively few submarine caves along the Levantine coast and there are few deep-sea rocky bottoms on report.

We have every reason to consider the Levantine sea to be an extremely stressed marine environment. In the past we emphasized the tropical, high-salinity, cul-de-sac situation of this basin, which is in contact with the world ocean only through the temperate Atlantic. We can now also add the extreme instability of its environments on the subrecent geological scale: active and ongoing tectonics, high sedimentation rates and last but not least, the extreme salinity fluctuations on the surface, induced by Nile run-off (Fig. 3.5) or Black Sea intrusions and the concomitant anoxic events in the deeper levels of the water column.

All this shows up in the considerable biotic depauperation of the Levantine basin, as compared with the more western Mediterranean basins. This is probably due in part to the present-day hydrographic and edaphic conditions and in part to the inherent inertia of the resettling process, possibly after the last "sapropelic event" (see above).

The benthos is most influenced by this combination of historic and actualistic circumstances. The zooplankton is not depauperated to a similar degree and successfully adapted its species inventory to the present-day oceanographic parameters.

Relative poverty characterized the Levantine basin for almost all the life-span of the Mediterranean and most certainly for the post-Messinian period of time. It had very few "positive" characteristics. In the last hundred years, the Lessepsian immigration from the Red Sea has probably for the first time initiated a biogeographic revolution in the Mediterranean: a phase of renewed "Tethys- like" contact with the tropical Indian Ocean. As with many other revolutions, this also started in a backyard, the Levantine basin.

3.3. Depauperation of the zoobenthos

For years, the reduced diversity in the Levantine Basin could be attributed to incomplete knowledge. Today this argument can be safely discarded.

Based on the careful census work on the Mediterranean benthos by Fredj (1974) and Fredj and Laubier (1985) one has today a more precise notion about the faunal diversity in this basin.

The present figure of 6,346 macrobenthic species is probably going to increase in the future, but it is evident that the Levant Basin is by far the poorest region of the sea. Ben Tuvia (1983) calculated that the Levant Basin has only 57% of the fishes of the "western Mediterranean", and 48% of the Mollusca; the figure for the less well-known Hydroidea is 26% and for the Polychaeta 38%.

The rate of impoverishment is probably smaller for the meiobenthos (based on a parallel calculation made by Por (1964a) for the Black Sea as compared to the Mediterranean) and also for the zooplankton (see below). We believe that a safe hypothetical figure would be a 30% impoverishment in the general diversity of the marine biota in the Levantine Basin. This might change in the future, even dramatically, because of the Lessepsian influx.

At a case-history level, the Levantine impoverishment is extremely evident: No species of the rich fauna of Gorgonaria of the western Mediterranean, including the noble coral *Corallium rubrum*, lives in the Levant. Among the decapod crustaceans the lobster *Homarus gammarus* and the spiny lobster *Palinurus elephas* do not live in our region.There are several more decapods missing, such as *Processa canaliculata* and several species of *Plesonika* (Holthuis, 1987). Among the cephalopods, a series of species of the small *Sepiola* are confined to the western Basin (Mangold and Boletzky, 1987). Even among the gastropods, there are some obvious absentees as for instance three species of the tritons, *Cymatium* (Gaillard, 1987). Peres (1958) discussed extensively the absence of molgulid and polyclinid Ascidiacea in the Levantine basin.

The very lively discussion about the Mediterranean endemism can be postponed for the moment. However, in Tortonese's important summing-up (1985), out of the 60 endemic Mediterranean fishes, only 3 are possibly limited to the "eastern Mediterranean" and there is no positive case of endemic echinoderms restricted to the Levantine basin. The stressing conditions and their lability over short time-spans did not allow for speciation in the essentially tributary Levantine basin.

A special aspect of the qualitative impoverishment of the Levantine basin is that of the "Levantine nannism". Older authors supplied many data indicating that the Levantine populations of common Mediterranean species are characterized by small body size. The nannism was discussed by Levi (1957) for sponges, Tortonese (1951) for fishes, Stephen (1958) for sipunculids and generally accepted by Peres (1967), Por (1978) and Sara (1985). Unfortunately, there are no recent studies of this interesting phenomenon and neither do we have a plausible hypothesis for it. The nannism might result from exceptional environmental factors, from the low productivity of the Levant or as a combination, as a peculiar form of r-strategy of the marginal populations of a species (Y. Ayal, pers. com.).

As far as the benthic associations are concerned, there are noteworthy differences between the Levantine Basin and the western Mediterranean. Some cases are extremely obvious: for instance the extremely important *Posidonia oceanica* meadows are not found in the Levant. Considering the enormous importance of this environment for the Mediterranean diversity (Peres, 1967; Ott, 1980; Ross et al. 1985 and many others), the absence of the *Posidonia* beds might be a principal factor in the Levantine depauperation. To our knowledge the easternmost appearance of this marine angiosperm is in the waters around Rhodes. Some authors consider that the place of the *Posidonia* meadows is taken in the "eastern Mediterranean" by the tropical and Indopacific seagrass *Halophila stipulacea*. This species appears more or less extensively only in the Aegean Sea, and there are no reports from the Levantine shores proper. Neither does the seagrass *Zostera* reach importance in the Levantine waters. Extensive vegetal cover on infralittoral bottoms is

Fig. 3.6. Cladocora caespitosa, Mediterranean zooxanthellate coral (photo D. Darom).

provided in the Levant only by the meadows of the green alga *Caulerpa scalpelliformis*, especially in the Bay of Haifa (Gilat-Gottlieb, 1959).

An infralittoral coralligenous community exists along the Levantine coasts (Gilat, 1964; Gorgy, 1966): this is primarily formed by calcareous red algae, sponges, bryozoans and serpulids. Though impoverished, it can certainly be equated with the western Mediterranean "coralligene". In the Bay of Haifa the coralligenous bottom around 10–12 m is built mainly by a frame of dead or vestigial "reefs" of the scleractinian coral *Cladocora caespitosa* (Fig. 3.6). On deeper bottoms, typical Mediterranean associations such as the gorgonaceous bottoms of *Isidella elongata* (Fredj and Laubier, 1985) and the "sponge bottoms" are not in evidence in the Levantine basin.

While in general the intertidal rocky fauna is similar to that of the western Mediterranean, there is, however, a very sharp decline in the diversity of the fauna as one proceeds southwards, along the Levantine coast and into the area of the Nile sediments. This relates to the type of rocky substrate available: south of the limestone rocks of Rosh Haniqra, many typical Mediterranean littoral species, like for instance the chiton *Middendorfia caprearum*, the sea urchin *Arbacia lixula* and the brittle star *Ophiura texturata* are absent or very rare. It seems that the sandstone substrate, quickly eroded, is not very suitable for the littoral fauna. According to Tsurnamal (1968) permanent abrasion by sand is detrimental for the development of sponges and other sessile organisms along the coasts of Israel. Last but not least, the season of low salinity following the Nile floods, especially during more wet, subrecent phases, should also be taken into account as a severe limitng factor for the littoral settlement.

It is evident that not only Egypt, but most of the Levantine basin has to be considered a daughter of the Nile.

The level bottoms of the Levantine basin present a different picture. Basically the fauna is similar to that of the western Mediterranean, especially starting with the depths of ± 30 m (Gilat, 1974): the indicator species are the "pan-mediterranean" *Pennatula rubra, Sabella pavonina, Aloidis gibba, Cardita aculeata, Leda pella, Cassidaria echinophora, Natica flammulata, Dentalium dentale, Echinocardium cordatum, Schizaster canaliferus, Bryssopsis lyrifera, Antedon mediterranea* and *Amphiura chiajei.*

The shallow infralittoral sands are devoid of the typical *Branchiostoma lanceolatum* community but on the other hand have some typically tropical modifications (see below).

Dexter (1988), comparatively analysed the sandy beach fauna (above 0.5 mm body size) of the Israel Mediterranean shore and the shores of the Gulf of Aqaba. The Mediterranean beaches were found to be almost barren, with only 7 species reported and 152 individuals, from a total sampling area of 1 square meter. The Red Sea beaches yielded from a 3.1 square meter total sampling area as much as 142 species and 4050 individuals. Dexter (op. cit.) however sampled only a small fraction of the meiobenthos which was retained by the sieves. Unfortunately, Dexter gives no comparative data with western Mediterranean sandy shores.

Considering the full diversity of the meiobenthic level-bottom fauna, I have the feeling based on the relatively good knowledge of the Copepoda Harpacticoidea (Por, 1964a) that there are no considerable differences if compared with the western Mediterranean. Still, it seems that at the level of the small fauna of the sediments, an important element is missing (possibly also in the whole Mediterranean): These are the "planktobenthic" calanoid copepods, cirolanid, and other isopods, hadziid amphipods and thermosbaenaceans known from other parts of the warm-ocean world (Caribbean, eastern Atlantic, Indian and Pacific oceans). They are giving rise to the fauna of submarine caves and crevices in an ongoing evolutionary process. Probably also missing are the alpheid, palaemonid and other small Decapoda Macrura which inhabit the dark overhangs, crevices and anchialine pools (see below) of the tropical belt. Despite the considerable tectonic uplifting in our area, especially during the Pleistocene, we have no "stranded" marine fauna (Stock, 1976). As mentioned above (p. 6) the subterranean crustaceans of marine origin found in our area are of a Miocene origin and the process of settling in the subterranean waters has stopped long ago: the species are "relics" with no contemporary kin in the open sea. As discussed by Por (1986), this is one of the consequences of the Messinian depauperation of the Mediterranean and perhaps especially of the Levantine Basin.

The muds of the bathyal and abyssal have hardly been investigated at all. The much discussed presence there of an elasipod holothurian *Kolga ludwigi* (= *Irpa ludwigi*) and of the ophiuroid *Pectinura vestita* (Marenzeller, 1895; Tortonese, 1985; Fredj and Laubier, 1985) is reported only from the abyssal samples collected in the Levant by the "Pola" in the past century. Reference has been made above to the statements by George and Menzies (1968), Menzies (1972) and following them, by Bacescu (1985), that the bathyal bottoms of the Levant are still inhospitable or even azoic after the last "sapropelic event".

3.4. Levantine zooplankton

The analysis of the Levantine zooplankton results in a slightly more complex picture. Furnestin (1979) gave an excellent summing-up of the faunal depauperation of the zooplankton in the eastern Mediterranean: Pteropoda decrease from 25 species in the west to 17 species in the east; Chaetognatha, from 15 to 12; Siphonophora from some 50 to 26; Hydromedusae from 70 to 24. Hyperiidea, decrease to 25% of their diversity. *Pyrosoma* does not appear in the eastern Basin and the Cyclosalpidae (Thaliacea) are reduced to one species. The depauperation seems to be relatively insignificant in the case of the Euphausiacea and the Appendicularia.

On the other hand however, there are in the Levantine basin other species of typically warm- and saline water plankton which are not found or are rather rare in the western Mediterranean.

Data are fairly complete for the planktonic copepods in the offshore waters of the Levantine Basin (Pasteur et al. 1976) and for the Israeli neritic waters (Berdugo and Kimor, 1968; Pasteur et al. 1976). Nearshore calanoid copepod communities were especially well studied near Beyrouth by Lakkis (1973, 1976, 1983). In general, the basic composition is the same in the Levant as in the rest of the Mediterranean. However, there are some differences. First of all, boreal species like *Calanus helgolandicus* do not reach our area. Second, there are differences in the predominance of species: this is for instance the case of *Paracalanus parvus* which is seen as extremely characteristic for the Levantine waters (Pasteur et al. 1976; Moraitou Apostolopoulou, 1985) or *Centropages violaceus* which is important in the Levant (Berdugo and Kimor, 1968). Some species are replaced by congeners, as in the case of *Centropages typicus* which is replaced by *C. kroyeri* in Lebanon (Gaudy, 1985). *Acartia negligens* is the most common species of this genus in the Levantine basin instead of *A. clausi* which is the more prevalent in the western Mediterranean.

There are other "Oriental" species in the Mediterranean plankton, which are predominantly found in the east: an example is *Sagitta serratodentata* (Furnestin, 1979; Casanova, 1986). There are similar cases among the Sergestidae (Cassanova, 1986) and the Globigerinidae (Herman, 1981).

According to Moraitou-Apostolopoulou (1985) the Levantine zooplankton is furthermore characterized by large populations of the pteropods *Cresseis acicula* and *Limacina inflata* and the appendicularian *Appendicularia sicula*. Rampal (1981) even considers the possibility that the Levantine basin might have functioned as a center of speciation. If so, this would be a quite exceptional situation.

In some cases, there is a possibility that some tropical species known till now from the Levantine Basin only, might be either circumtropical expatriates or Red Sea immigrants. Such may be the case of the calanoids *Paracalanus aculeatus*, *Calocalanus pavo* (Moraitou Apostolopoulou, 1985), *Pontellina plumata* and *Labidocera detruncata* (Lakkis, 1976).

The influence of the Suez Canal connection as well as the impact of the Aswan dam on the Levantine zooplankton will be discussed below.

3.5. Positive features of the Levantine biota

It would be unbalanced to emphasize only the negative features of the Levantine marine world, even if these features are so prevalent. "Positive" features are found not only among the zooplanktonic species (see above); there are some "autochthonous" warmwater biota typically found in the Levantine basin prior to the recent Lessepsian influx.

The warmwater biota are as a rule encountered also in the southern Mediterranean, i.e. the Sidra gulfs.

In a previous chapter we mentioned the warmwater "Tyrrhenian" fauna which characterized several warm Interglacials in the Mediterranean. This is a faunal complex also called *Strombus bubonius* fauna, Senegalian fauna or Eutyrrhenian fauna (see Peres, 1985). Much of this fauna, presently inhabiting the tropical west African coast, left its fossil marks in the Pleistocene Mediterranean. However, some of the Senegalian mollusks still live in the warmwater corner of the Mediterranean: such are *Fissurella nubecula, Pirenella conica, Cypraea lurida, Purpura haemastoma, Mitra fusca* and *Clavatula nifat* (Peres, 1985). To this component, Peres adds several ascidians. A few genera of fishes like *Epinephelus, Serranus* and *Crenilabrus* are also Senegalian elements. Holthuis and Gottlieb (1958) in their list of Decapoda from the Israeli coast, mention the following species which might be considered "Tyrrhenian relics" in the Levantine basin: *Athanas amazone, Salmoneus jarli, Micropanope rufopunctata, Maja goltziana, Albunea carabus, Ocypode cursor* and *Pachygrapsus transversus*. The two last species are extremely abundant and widespread along the Levantine and Libyan-Egyptian coasts (see also discussion on this point in Peres, 1985). Among the Levantine algae, there are no doubt also a number of Senegalian expatriates (Bacescu, 1985).

At the population level, the Levantine basin has some definite tropical characteristics. Gilat (1964, 1974) describes from the Israeli shores a typical "crab community" (sensu Thorson, 1957) of the shallow sandy infralittoral, in which

Fig. 3.7. Vermetus platforms along the Israeli coast (photo D. Darom).

Diogenes pugilator, Macropipus pusillus and *Portunus hastatus* are preeminent species.

A positive feature of the Levantine shores is the extremely marked development of the vermetid trottoirs (Fig. 3.7), built by *Vermetus triqueter* and *Dendropoma petraeum*. These are in a sense a community which parallels the western Mediterranean "trottoirs" of the red alga *Lithophyllum tortuosum*. This association was found in subfossil stage in Rhodes (Laborel, 1981). The vermetid platforms are especially well developed on the kurkar sandstones of the Israeli and southern Lebanese shores (Safriel, 1975) and tend to be less in evidence on the Syrian coast. Being an association characteristic for the uppermost infralittoral, the "Infralittoral Fringe" of the authors, this well-developed formation does not harbor a rich fauna.

Considering that the *Vermetus* bioherms are well developed also in the Canary Islands, the Bermudas and the Island of Fernando de Noronha, it may appear that they are subtropical Atlantic parallels to the tropical coral reefs. Tzur and Safriel (1978) suppose that the well-developed *Vermetus* platforms of the Levantine coasts are rather to be explained by the presence of the soft eolianitic base rock and to the gradual uplifting of the shore, which compensates for the erosion.

3.6. The progress of the Lessepsian migrants

Without doubt, the most important biogeographic phenomenon witnessed in the contemporary oceans is presently unfolding in the Levantine Basin of the Mediterranean. This is the invasion and settling of hundreds of species of Indo-Westpacific origin in this sea. These organisms have been called "Lessepsian migrants" (Por, 1969b), a name which has received wide acceptance. Lessepsian migrants are Red Sea species which used the opportunity created by the opening of the Suez Canal in 1869 in order to gradually advance and colonize the Mediterranean. A considerable body of literature deals with this subject and has been reviewed and discussed till 1976 by Por (1978). In that treatment of the issue extreme care was taken to cut down the number of Lessepsian migrants to those obvious cases where the animals are present in the Suez Canal and at the two ends of it, preferably also showing a documented gradual expansion. This left aside the tropical or even circumtropical species of the eastern Mediterranean which either colonized through the Gibraltar (for instance "Senegalian" elements) or reached the Mediterranean from the Red Sea earlier in the history ("Prelessepsian" elements). Species introduced passively to the Mediterranean by ships passing through the Suez Canal, were not considered as trustworthy Lessepsian migrants. Por (op. cit.) therefore lists "High Probability Lessepsian migrants" (128 species) separately from "Low Probability Lessepsian migrants" (76 species) and rejects 33 species which have been claimed previously to be migrants. The caution taken in separating the two first categories has no meaning in our present context: they both contain Indopacific newcomers in the Levantine basin. Moreover, in the 12 years which passed since the above-mentioned compilation has been made, some of the Low-Probability migrants "upgraded" their status.

The Lessepsian migration is still unabatedly going on. Por (1978) considered that the influx was nearing a plateau: there are no clear signs for this yet. The Suez Canal, closed for navigation between 1967–1975 (and even obstructed by a dam between 1973–1974) has been deepened and widened, the cessation of the Nile flow to the Mediterranean created more stenohaline conditions at the Canal outlet: yet there was no increase in the volume of the migration either. The filter action played by the Gulf of Suez and the Canal itself, did not decrease.

As a consequence, the main features of the Lessepsian migration remained the same: first of all, there was no penetration of reef-forming coelenterates and their associated fauna. The recently reported reef-forming coral which settled in the Gulf of Genua around 1966 and spread to various west Mediterranean localities, is the south Atlantic *Oculina patagonica* (Zibrowius and Ramos, 1983) and not an Indopacific one. The Mediterranean (and Levantine) coral *Cladocora caespitosa* already mentioned above, is an autochthonous zooxanthellae-bearing species, but it forms only marginal populations or residual subfossil reefs considered by Zibrowius (1980) to be of Tyrrhenian age. Schumacher and Zibrowius (1985) consider *Cladocora* to be "rather close, ecologically, to tropical reef-building corals".

Fig. 3.8. Lessepsian migrant Mollusca Pelecypoda: 1. *Pinctada radiata*; 2. *Malleus regula*; 3. *Brachidontes variabilis*; 4. *Paphia textile*; 5. *Clementia papyracea*; 6. *Mactra olorina* (from Por, 1978).

Till now, none of the many zooxanthellae bearing organisms of the Red Sea could get a foothold in the Mediterranean and therefore no hermatypic reef building activity seems plausible in the near future. Repeated reports of the zooxanthellate bottom dwelling medusa *Cassiopeia andromeda* in the Mediterranean (Spanier, pers.comm.) have to be substantiated.

There was no increase in the number of holoplanktonic Red Sea species in the

Fig. 3.9. Lessepsian migrant Mollusca Gastropoda: 1. *Rissoina bertholetti*; 2. *Cerithium scabridum*; 3. *Rhinoclavis kochi*; 4. *Thais carinifera*; 5. *Murex tribulus* (from Por, 1978).

Levantine plankton. The type of settlers remained essentially the same: demersal and benthic species with relatively broad spectra of substrate requirements and a wide range of thermo-saline tolerance.

The last decade did not advance our knowledge about the Suez Canal biota themselves. The new records of Lessepsian migrants from the Mediterranean itself have been the result of occasional observations, and there was no concentrated effort to follow the deployment of this unique biogeographical phenomenon (like the project "Biota of the Red Sea and eastern Mediterranean", between 1967–1973 (Por et al., 1972)).

There are some recent states of art of the Lessepsian migration in several taxa. Barash and Danin (1986) set the number of Lessepsian migrant mollusks which can be considered resident species at 44. They add a further 47 species considered as doubtful, since they were collected only as isolated shells. Since such isolated and therefore not sufficiently reliable reports are usually confirmed later, one can assume that there are today some 90 Indopacific mollusk species in the Mediterranean. It is worth mentioning that Barash and Danin's listing contains for the first time 5 species of the ectoparasitic Pyramidellidae (see also Aartsen, 1983 and Aartsen and Carrozza, 1979) and that the number of Lessepsian Opisthobranchia

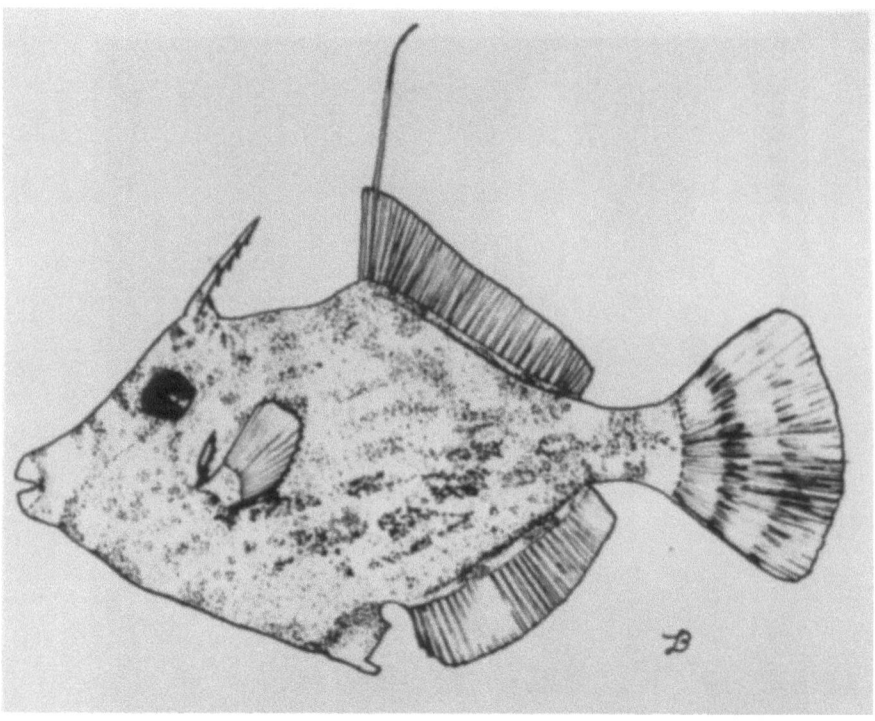

Fig. 3.10. The Lessepsian migrant *Stephanolepis diaspros* (from Ben-Tuvia, 1976).

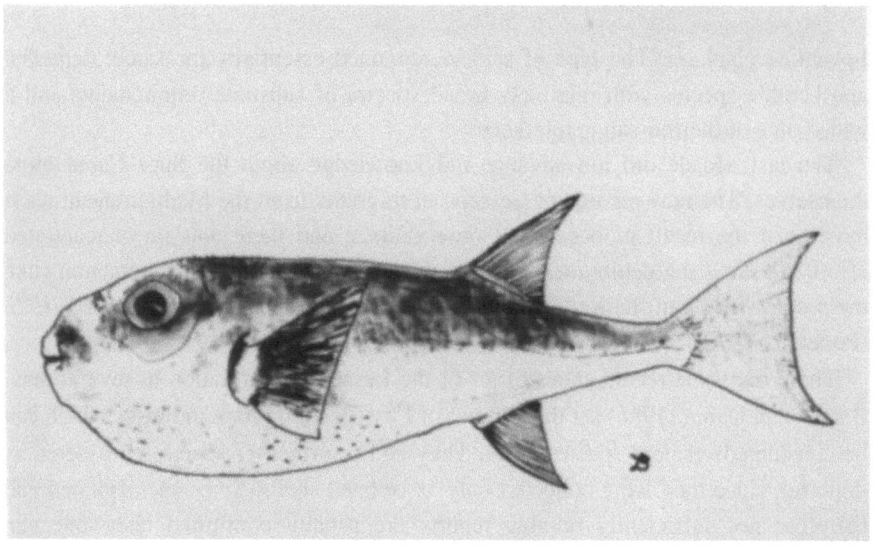

Fig. 3.11. The Lessepsian migrant *Sphaeroides spadiceus* (from Ben-Tuvia, 1976).

increased to 6. Compared to Barash and Danin's last census (1977), the number of Lessepsian mollusks increased by 29 species (Figs. 3.8 and 3.9).

Ben-Tuvia (1985) presents an increased list of migrant fishes: 41 species, compared with the 35 species mentioned a few years ago (Ben-Tuvia, 1983) (Figs. 3.10 and 3.11). The shark *Carcharhinus brevipinna* is considered now by Ben Tuvia to be a circumtropical species and therefore taken off the list.

The new immigrants are three species which are already forming large populations: *Sillago sihama* (representing another new family in the Mediterranean, the Sillaginidae), *Pempheris vanicolensis* (the new family Pempheridae) and the gobiid *Oxyurichthys papuensis*. Further species: *Lutjanus argentimaculatus, Rachycentron canadum, Muraenesox cinereus* and *Arius thalassinus* are new reports based on single specimens. Another species of new Lessepsian migrants is added by Golani (1987) namely *Torquigener flavimaculosus*, a second migrant pufferfish. As to the large and peculiar cobia, *Rachycentron canadum* (Golani and Ben-Tuvia, 1986) it is still too early to decide whether it is a migrant or a circumatlantic fish.

Two recent additions, that of the goby *Silhouettea aegyptica*, a previously misidentified species, an that of the trunkfish *Tetrosomus gibbosus* (Spanier and Goren, 1988) bring the number of the migrant fishes already to 44 species. Lessepsian migration added 14 new families to the ichthyofauna of the Mediterranean.

Ben-Eliahu (in press) lists 7 Lessepsian migrant serpulid polychaetes. Six of these belong to genus *Hydroides* which can be transported on ship hulls. *Spirobranchus tetraceros* is by now present as far as the Island of Rhodes (Ben Eliahu, op. cit.). This is a considerable numerical increase compared with the 2 serpulids listed by Por (1978). The decapod crustaceans, which together with the Pisces and the Mollusca are a "leading" taxon among the migrants, have not been intensively studied of late. However, Galil et al. (in press) report the appearance of *Matuta banksi* and the presence of *Metapenaeopsis aegyptiacus* on the coasts of Israel. Almaca (1985) mentions an additional xanthid crab among the Lessepsian migrants, namely *Sphaerozius nitidus*, not contained in previous lists.

Cherbonnier (1986) identifies a first Lessepsian holothurian in the Levant basin, from museum material collected along the Israeli coast and from Cyprus, about 20 years ago.

Recent reports do no indicate any important increase in the number of Lessepsian migrants among the zooplankton. Lakkis (1976) lists at least two clear-cut Lessepsian migrants among the calanoid copepods: *Labidocera madurae* and *Acartia fossae*. Other reported cases are eventually circumtropical elements (see above). Moraitou Apostolopoulou (1985) mentions again the issue of gene-flow enrichment of pre-existent tropical species, through the Suez Canal. However, unless studied in the detailed way in which similar genetic exchanges have been studied for *Calanus helgolandicus* and its Black Sea populations (Fleminger and Hülsemann, 1987), such statements are merely hypothetical.

Though not a recent report, one has to add a first arrow worm *Sagitta neglecta* to

Fig. 3.12. Asterina burtoni, a succcessful Lessepsian migrant (original).

the list of Lessepsian migrants (Guergues and Halim, 1973): this species has been previously reported from the Suez Canal and the two authors reported it from Alexandria, in the open Mediterranean.

Furnestin (1979) was very prudent on the subject of the Lessepsian zooplankton species. She wrote: "One should only retain the more recent records of Indo-Pacific species which were observed at the same time in the Red Sea, the canal and various places of the Levantine basin.." This coincides with the views of Por (1978). However, Furnestin was very optimistic concerning an increased planktonic migration following the hydrographic changes in the Suez Canal and in the Levantine Basin. The future will show if she was right.

There are some more recent discussion papers about the mechanism, the success and the predominantly unidirectional faunal movement through the Suez Canal. Vermeij (1978) reached conclusions very similar to those by Por (1978): faunal depauperation of the Levantine basin and saline preadaptation of the Red Sea species to Suez Canal-type salinities are the main reasons for the prevalent Lessepsian migration. Agur and Safriel (1981) consider on the basis of a "mathematical- hydraulic" model that Red Sea species have 3 times more chances to enter the Suez Canal. For those which successfully breed in the metahaline Bitter Lakes, the chances to cross the Suez Canal are 75 times higher than for those which cannot establish themselves in these lakes. Following a series of detailed studies about migrant versus non-migrant mollusks, Safriel and Ritte (1985) reach the surprising conclusion that the Suez Canal does not act as a "genetic bottleneck" and that conspecific populations on both sides of the Canal have basically the same allelic structure. What is probably important according to these authors is the basic reproductive strategy of the species: r-strategist species are prefered since their high intrinsic growth rate at low densities reduces the dangers facing a small colonizing population. This interesting discussion is based on data collected by the group of Safriel on the biology of two species pairs: the Lessepsian migrants *Cerithium scabridum* and *Brachidontes variabilis* versus the non-migrant Red Sea species *Cerithium caeruleum* and *Modiolus auriculatus*.

Quite recently, Golani (1988) published a monographic analysis of the successful migration of two species of goatfish: *Upeneus molucensis* and *U. asymmetricus*. He compared them both with the other non-migrant Mullidae in the Gulf of Aqaba as well as with the two autochthonous Mediterranean species (*Mullus barbatus* and *M. surmuletus*). It was found that the two migrant species are typical sandy bottom inhabitants, whereas the non-migrants are at least in their adult stage connected with the coral reef. Among the sandy bottom fishes of the Gulf of Aqaba, Golani (op. cit.) found about 24% of migrant species, whereas for the totality of the fishes of the Red Sea, the percentage is only 4.3%. Although the principal migrant goatfish is an r-strategist, Golani does not give an absolute value to this adaptation. Like in the above mentioned mollusks, the recently settled Mediterranean populations of the two species of *Upeneus* do not show genetically tangible differences from their parent populations.

It is now evident that the Lessepsian migrants occupy preferentially the shallower infralittoral levels. They are indeed rare in the intertidal zone as well as at depths greater than 50 m (Gilat, 1974; Tom, 1976; Galil and Lewinsohn, 1981; Ben-Tuvia, 1985). The Lessepsian migrant species *Rhinoclavis* (= *Cerithium*) *kochi*, *Charybdis longicollis* and *Oratosquilla massawensis* are massive and characteristic inhabitants of the sandy-muds at 20–30 m depths.

As far as the percentual importance of the migrants among the Levantine species is concerned, the following figures are now available: Decapod crustaceans, 20% (Galil, 1986); fishes, 12% (Ben-Tuvia, 1983; mollusks 9.4% (Barash and Danin, 1986); polychaetes, 11% (Ben-Eliahu, in press). Even among the 177 decapods of Turkey, listed by Kocatas (1981), 18 are considered to be of Indopacific origin. The assumption expressed by Por (1978) that the Lessepsian migrants constitute 10% of the total species diversity in the Levant Basin, seems to have been a good approximation.

The impact on the fisheries of the Israeli coast increased. Ben-Tuvia (1985) considers that during the last 10 years the proportional importance of the Red Sea immigrants increased from 21% to 33.6% in the trawl fisheries. Summarizing the proportions in the pelagic and the inshore fisheries, the overall proportion is of 16.2%. It is interesting also to quote Galil (1986) according to whom 5 of the 7 species of penaeid shrimps caught along the Israeli coasts are Lessepsian migrants.

Taking into account this level of Indopacific influence and especially the appearance of many genera and families new to the Mediterranean, it is indeed justified to consider the appearance of a new biogeographical province within this sea. The proposal to recognize a "Lessepsian Province" (Por, 1983a) within the Mediterranean has been generally accepted by the authors. This province will probably contain, besides the Levantine Basin, also the Ionian Sea and especially the southern Mediterranean; the Aegean Sea, and especially its northern part, will not be part of this new province (Fig. 3.13).

As mentioned already, we have the feeling that only the presence of zooxanthellae-bearing organisms (both benthic and planktonic) will turn the Levantine basin into a tropical province. It is more probable (Por, 1983a) that the

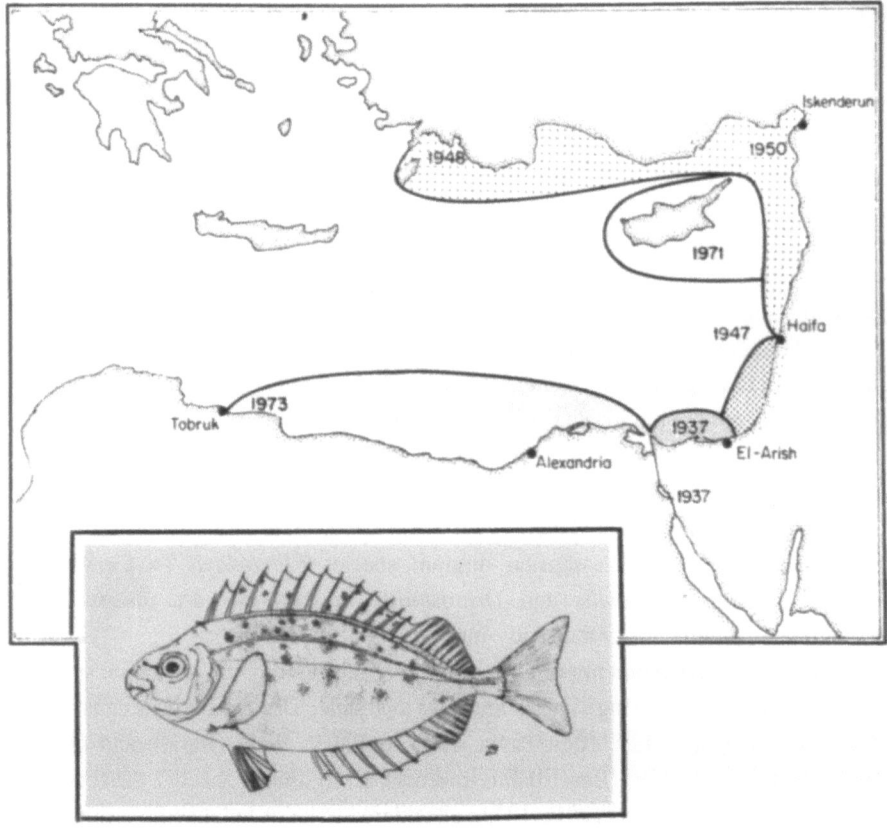

Fig. 3.13. The gradual advance of the migrant fish *Siganus rivulatus* (inset) in the Mediterranean (from Por, 1978).

new province will have to be considered similar to the subtropical provinces which are interposed between the tropical and the warm-temperate provinces in other parts of the world ocean.

As mentioned above, the settlement of Lessepsian migrants in the Mediterranean is still going on. Ben-Tuvia (1985) confidently assumes that within the next 50 years the number of immigrant fishes will increase to about 80 species; out of a total of 350 species in the Levant basin, this will represent about 30%.

3.7. The impact of the Aswan High Dam

A second engineering project is deeply influencing the Levant basin. This is the tapering off of the Nile outflow following the building and closing of the Aswan High Dam in 1967. The impact of the near cessation of the Nile inflow is probably going to influence the oceanographic parameters of the Levantine basin as a whole and not necessarily the "qualitative" biogeographic characteristics of this basin. In this, the impact will be totally different from that of the 100 years older Suez Canal.

The Aswan dam issue has been extensively discussed and investigated (see among others, Oren, 1969, 1970; Aleem, 1969; Hammerton, 1972; and lately Mancy, 1981). The crux of the phenomenon is the fact that most of the yearly 84 billion cubic meters of nutrient-loaded freshwaters do not reach the Levant basin anymore. This is partly compensated by the increased outflow of used water from the towns and fields of the Nile Delta (Khalil et al. 1983). Furthermore, about 81×10 cubic meters of silty sediment does not add to the Nile cone and to the Deltaic and Levantine shores.

The event, though abrupt and man-made, is not a new one. As shown above (pp. 30–31) the Nile was probably reduced more than once by climatic changes to the level of a huge intermittent wadi. This is now fairly convincingly documented for the period before about 12,000 years B.P. Also, as shown by Hammerton (1972), the Nile already suffered a series of man-made, though undoubtedly smaller, engineering impacts.

Our assumption is that the Aswan event will start a compensative fluctuation in the oceanographic conditions of the Levant basin, basically similar to the ones that occurred in the past.

As shown by Aleem (1969) and Halim (1976) the phytoplankton concentrations along the Egyptian coast dropped to 10% after the Nile closure. Similar figures are given for the zooplankton biomass by Dowidar and El Maghrabi (1971) and El Maghrabi and Dowidar (1973) for the area of Alexandria. Moreover, there was a modification of the annual dynamics: the autumn peak of phyto- and zooplankton biomass following the Nile floods is less pronounced or is even exceeded by a December peak. But basically the bimodal type of dynamics, with two annual peaks remained (see Halim et al. 1983; Khalil et al., 1983). According to Dowidar (1984), the importance of the winter peak even increased of recent.

Berman et al. (1986) still find higher values of chlorophyll standing stock near the Egyptian coast than along the Israel coast; however, there is no way to compare with Pre-Aswan chlorophyll determinations. Dowidar (1984) calculates a present primary production of 55.5 gC m^{-2} y^{-1} along the Egyptian coast, as compared to 36 gC m^{-2} y^{-1} along the Israeli coast.

The most impressive negative result of the Aswan dam closure was seen in the disappearance of the huge concentrations of *Sardinella aurita* in front of the Nile openings, during the late-summer and fall season. According to El Hehyawi (1974) the annual catch of this fish was of 18,000 tons in 1962 and fell to 460 tons in 1968. But according to this author there has been a recovery to more than 6000 tons during 1979 and 1980. This would be due to an increased runoff of nutrient-rich freshwater reaching the Mediterranean: after a minimum of one tenth of the original Nile discharge in the first years of the Aswan Dam filling, it stands now at one quarter of the original discharge. Moreover, of the present annual discharge of 12 billion cubic meters, only 3 billion is Nile water while the rest is agricultural runoff (Dowidar, 1984).

There has been no influence on other fisheries and it is even difficult to prove that the overall population of *Sardinella* in the Levant Basin has been prejudiced (Bebars, 1981). In the delta lagoons the closure of Aswan did not have negative influence on the fish yields (Bebars and Lasserre, 1983).

The cessation of the Nile flood changed the regime of the currents in the northern Suez Canal (Morcos, 1967; El Sabh, 1969). The fact that during the flood months there is no fresh-or brackish water barrier anymore on the northward progress of the potential Lessepsian migrants, should have facilitated the migration. We have no ways to prove that this has indeed happened. Neither do we have the proof that the disappearance of the low-salinity barrier in front of the Nile delta did indeed open a westward road to the Lessepsian migrants. The fact that after Aswan the reports of Lessepsian migrants from the north African shores and from Malta and Sicily have multiplied can be due to increased research or even only to a normal and gradual process of expansion.

4. Northern Red Sea and the Gulfs

4.1. The uniqueness of the Red Sea

There is no sea of the world ocean which compares to the Red Sea. It is the youngest oceanic body, not more than 20 million years old; in the early Pliocene, it changed its connections through the uplifting of the Isthmus of Suez and from a satellite gulf of the Mediterranean it became a tongue of the Indian Ocean; moreover, through the time-lens of the geologists it is still ripping up from south to north. Being an exceptional sea which does not receive any permanent freshwater inflow and situated in an extremely arid area of the globe, the Red Sea is a metahaline sea. By the same token, the Baltic and Black Seas which receive excessive run-off are brackish seas. In the northern Red Sea, evaporation values reach more than 1 cm per day.

The salinity of the open Red Sea, an average 41 ppt, reaches even some 42.5 ppt in the open waters of the Gulf of Suez. In this it resembles the Persian Gulf. But unlike the shallow Persian Gulf, the deep Red Sea remained a marine waterbody even during the low Pleistocene sea levels, though experiencing even higher salinity values than today (see above).

This is a deep and extremely narrow sea, nearly 2000 km long and and on average 200 km wide. The average depth is 491 m and the maximal depth 2850 m (Head, 1987). Moreover, its northern gulf of Aqaba is only 20 km wide and over 1830 m deep; it is the narrowest oceanic chasm of the world (Por and Lerner-Seggev, 1966). Adding to this the data obtained from magnetometric and gravimetric measurements, it results that the northern part of the Gulf of Aqaba represents a 10 km deep fissure in the continental crust (Ben Avraham, pers. comm.).

The relatively deep Red Sea has a very aberrant regime of temperatures: they do not decrease below 21–22°C, even on the deepest bottoms and the thermocline is consequently very weakly developed.

In a sense, the Red Sea is similar to the Mediterranean: it is a sill-confined deep oceanic body with limited exchange with the nearest open ocean. As a consequence, there are clear hydrographic gradients from the opening to the blind end of the water body. In the Mediterranean, because of its location, there is an increasing salinity and temperature gradient from west to east; in the Red Sea the south to north gradient is of increasing salinity but of decreasing temperature.

Because of the complicated topography and local river and/or cold water influxes, the Mediterranean is a complex structure of basins (see above). The Red Sea on the other hand is an oversized experimental tank, almost linear and with practically no topographic obstructions, no bays and no large islands.

If it were better known, the Red Sea would be a laboratory for the study of the gradual adaptation, or cyclic oscillations, of marine biota vis-à-vis increased salinity, decreasing temperatures and declining biological productivity.

The Red Sea is, however, one of the least known oceanic waters. It is symptomatic, that there was only one systematic scientific cruise with the Red Sea as its main object: the Austrian "Pola" Expedition of 1895–1896; 1897–1898. The hydrographic data collected then and analyzed by Luksch (1898) and Natterer (1898) are still widely used and discussed; the deep-sea collections of the "Pola" (see Sturany, 1899; 1903) were for almost a century, the only source of information in this respect. Thus it was possible that on board the "Mabahiss" which visited the Gulf of Aqaba fifty years after the "Pola" was there, Crossland (1939) could ask himself whether this "most desolate sea in the world" has any coral reefs at all!

The first expedition aimed at the study of the Gulf of Aqaba, was the "Manihine" expedition in the winter of 1948–1949. Almost all additional knowledge about the open Red Sea, is also post-World War II. In fact, the renewed oceanographic research started almost by accident, when in 1963, on the way to the Indian Ocean, the "Discovery" came upon the first site of hot deep-sea brines in the Red Sea. This discovery, the first of its kind, brought the Red Sea into the center of the oceanographic interest.

The area of the brines has been intensively studied since 1970 by the Saudi Sudanese Red Sea Joint Commission (SSRJC) (Head, 1987) and the German vessels "Sonne" and "Valdivia".

There was much pioneering zoological research on the Red Sea shores, starting with Forskal (1761–1767) and continued with Hemprich, Ehrenberg, Rüppell and Klunzinger. For some time the reefs of At Tur on the Sinai coast were "the reefs" for the new darwinian biologists: Ernst Haeckel visited them in 1876. But only after the establishment of the El Ghardaqa, Elat, Aqaba, Port Sudan and Jeddah laboratories did the on-site modern research start. Until 1956, there was not even any clear knowledge as to how far north the mangrove forests thrive along the coasts of the Red Sea.

The research activity of the last decades concentrated more in the field of geology and prospection of the mining possibilites around the hot brines areas. Marine biology followed as a byproduct. Therefore we are today still largely ignorant about the deep-sea fauna, the plankton distribution and the meiobenthos of the Red Sea. The coral reefs and their associated fauna, primarily fish, mollusks and echinoderms are fairly well known. But as seen from a map prepared by Mergner (1984), a broad belt on both sides of the Tropic of Cancer: nearly 1000 km of the Red Sea from Al Ghardaqa to Jeddah are very poorly known. Extremely little knowledge has been gained about the Gulf of Suez and taking into account the massive pollution of this area, much of the natural conditions might already be radically impaired.

4.2. Northern Red Sea and its metahaline past

There is no doubt that a treatment of the Levantine marine biota has to include the two northern gulfs of the Red Sea: the Gulf of Aqaba (or Gulf of Elat) and the Gulf of Suez. These are two well-defined waterbodies. However, since they are interconnected through the open Red Sea, it is methodologically impossible not to deal with the northern part of this sea. The problem is: how far south should we set the limit of our interest?

The southern Red Sea, south of approximately 20° N is situated in the climatic belt of the trade winds, whereas the northern Red Sea has a different climatic regime. Whereas the northern part has a certain hydrography of its own, the southern part is dominated by the inflow from the adjacent Gulf of Aden, at least during the southeast monsoon, i.e. roughly during the winter months. According to Neumann and McGill (1967), the influx of Indian Ocean waters is felt even after the onset of the summer northwest monsoon, i.e against the wind direction; the reason is the accumulated water deficit in the desertic northern Red Sea. The limit

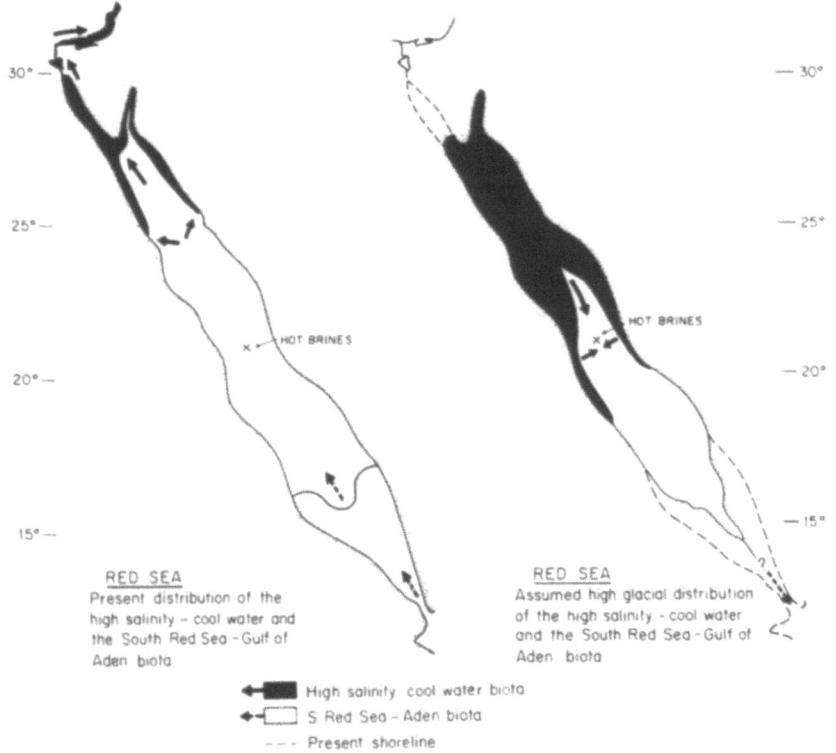

Fig. 4.1. Glacial and Interglacial pulsations in the Red Sea (from Por, 1978).

of this seasonal influx of relatively low-salinity water (37–38 ppt) with high productivity and loaded with Indian Ocean phyto- and zooplankton is as a rule situated between 15–20° N. In the same approximate area several other changes are occurring. On the line Port-Sudan-Jeddah (i.e. around 20° N) there is a nodal area of the tidal fluctuations (Morcos, 1970): The incoming Indian Ocean tide decreases in height from Bab el Mandeb until it reaches a practically "no-tide" situation; northward there is again a gradual increase of the tidal fluctuation, this time an independent Red Sea tide (see also Edwards, 1987).

The geologists clearly distinguish the area of the northern Red Sea, where the reduced and even non-existent magnetic anomaly indicates a young stage of rifting and where the deep-sea basalts are still covered with evaporitic salts (Girdler, 1984); as a consequence an axial trough is still absent there (Bonatti, et al., 1984). South of 20–24° N, there are strong magnetic anomalies and an axial valley of deep-sea basalts is already present. The area which separates the northern from the southern Red Sea also contains the known concentration of the hot brine spots which received much attention during the last two decades (see Degens and Ross, 1969). This area, to complicate things, is recently being called "Central Red Sea".

Considering now the Pleistocene past of the Red Sea, the two segments had different histories. This is a consequence of the cyclical "pulsations" of the glaciations and of their influence on this sea. As schematized by Por (1975a; 1978), during the low eustatic sea levels of the glacials, the northern Red Sea had very high salinity values and low temperatures (Fig. 4.1). Following older authors, Head (1987) considers that the salinity values were hypersaline in the Red Sea during its total or partial isolation from the Indian Ocean. According to him it was "like the present Dead Sea". This was not the case at least in the Gulf of Aqaba and the northernmost Red Sea. Reiss et al. (1980) found there during the last Glacial salinity values of slightly over 50 ppt and temperatures of about 17°C (as compared to 41 ppt and 21°C today).

The northern part of the Red Sea is characterized by the presence of the hardy survivors of the extreme hydrographic conditions of the last (and previous) Glacial; the southern Red Sea is characterized by the gradual predominance of the open-Indian Ocean biota, which reinvaded the sea, after each improved oceanic contact.

The survival of marine life in the Glacial Red Sea can be explained by the fact that "metahaline" adaptations evolved there over time. The metahaline biota will be discussed in more detail below (pp. 90–99). Here it is enough to mention that despite salinities as high as 60–70 ppt, a reduced diversity of marine biota can still be found in the lagoons of the northern Red Sea. There is every reason to assume that these biota could have been repeated survivors of the Glacial salinity crises which this sea underwent.

Price (1982) reports populations of echinoderms which in the Persian Gulf are found at exceptionally high salinities, although with stunted specimens. Two of them, *Astropecten polyacanthus* and *Asterina burtoni* (see Fig. 3.12), both also known in the Red Sea, survive in lagoons with a salinity of 60 ppt.

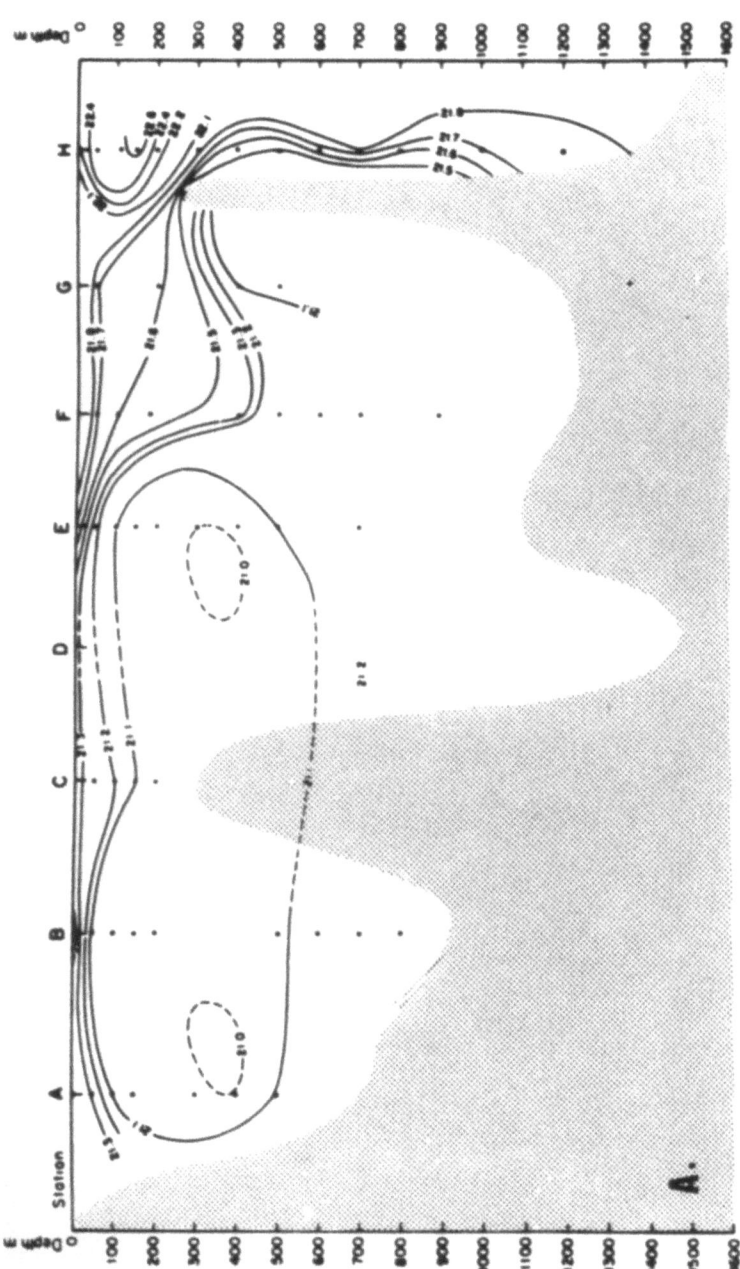

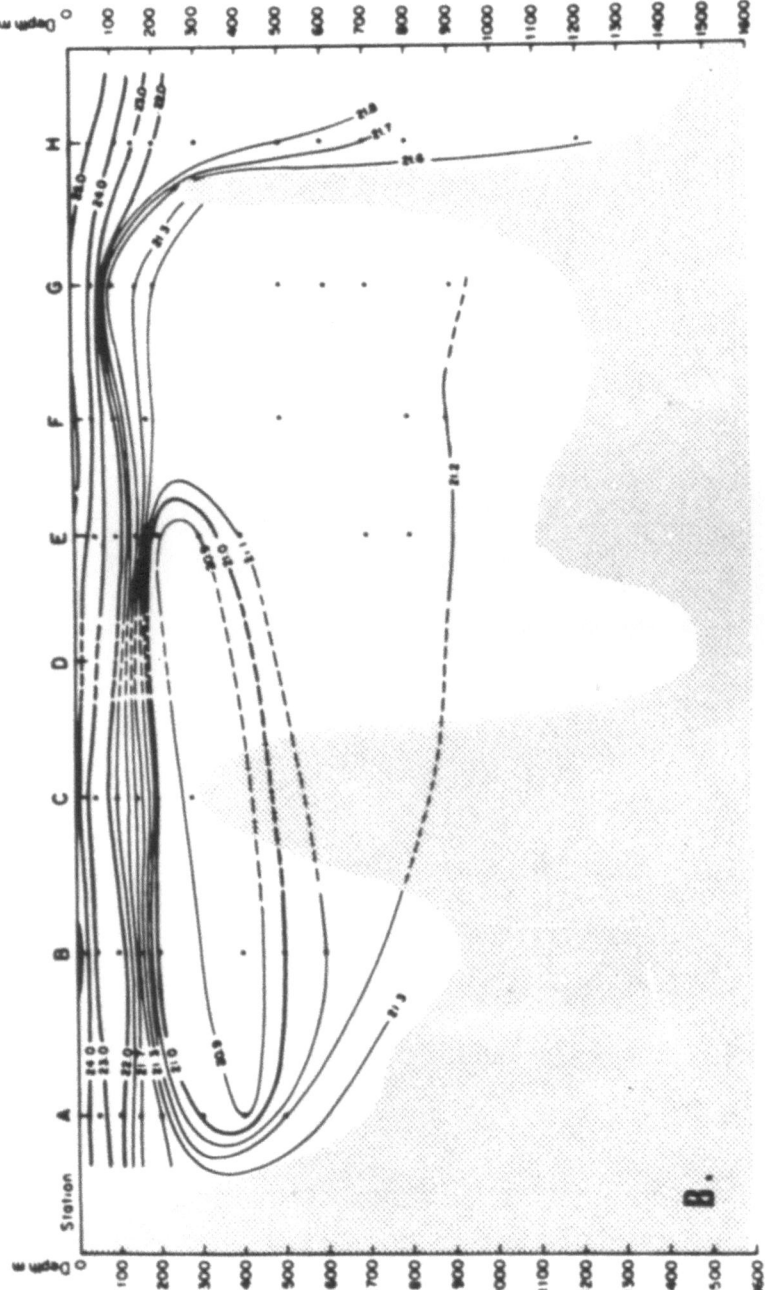

Fig. 4.2. Isotherms (°C) in the Gulf of Aqaba. (A) January–March and (B) April–October. Stations are indicated: A – Northern Gulf; G – Straits of Tiran; H – Open Red Sea (from Reiss and Hottinger, 1984).

4.3. Two Gulfs – two non-identical twins

The two gulfs bathing the Sinai Peninsula are identical at first glance but completely different in most respects. The Gulf of Aqaba, also called Gulf of Eilat or Elat (the Elanitic Bossom of the classical geographers) is a miniature Red Sea, a continuation of the Red Sea Rift, its youngest, active, and very deep branch (1830 m maximum depth). The Gulf is separated from the Red Sea by the shallow 252 m deep Tiran sill. The Gulf of Aqaba is nearly as deep as the Red Sea; however, its width is about 10 times less (Por and Lerner Seggev, 1966). There is hardly any horizontal N-S salinity or temperature gradient in this deep gulf. In all, the temperature of the Gulf is only about 0.5°C lower than that of the open Red Sea; salinity is equal to the northern part of it. The Gulf of Aqaba has a thermohaline seasonality of its own: during most of the year (April-December) the gulf waters are stratified, though difference in temperature between surface and deep bottoms is only in the range of 6°C (27–21°C) and there is practically no salinity gradient; during the rest of the year (mainly January-March) the water column is also homothermic in most of the Gulf, with the exception the southernmost corner (Klinker et al., 1976). This is quite different from the main body of the Red Sea, which has a permanent thermal stratification (Figs. 4.2. and 4.3).

The bottomward decrease in dissolved oxygen is very gradual and there are still nearly 4 mm/l O_2 on the deep bottoms. This situation is also quite unlike the open Red Sea: There an oxygen minimum layer builds up below 100 m and concentration falls to 0.3-0.5 ml/l around 400 m depth (Weikert, 1982, 1987); the dissolved oxygen content then increases slowly again to around 2 ml/l on the deep bottoms (Fig. 4.4).

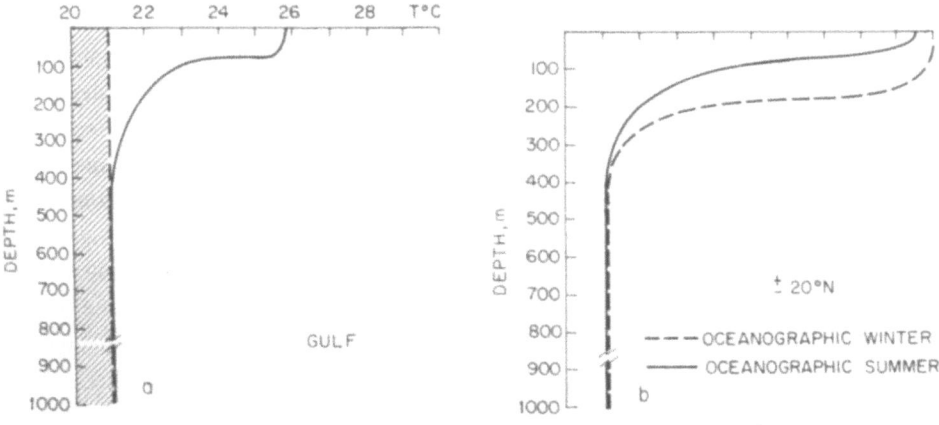

Fig. 4.3. Simplified temperature distribution a) in the Gulf of Aqaba, b) in the central Red Sea shaded column emphasizes homothermy (after Reiss and Hottinger, 1984 and Weikert, 1987).

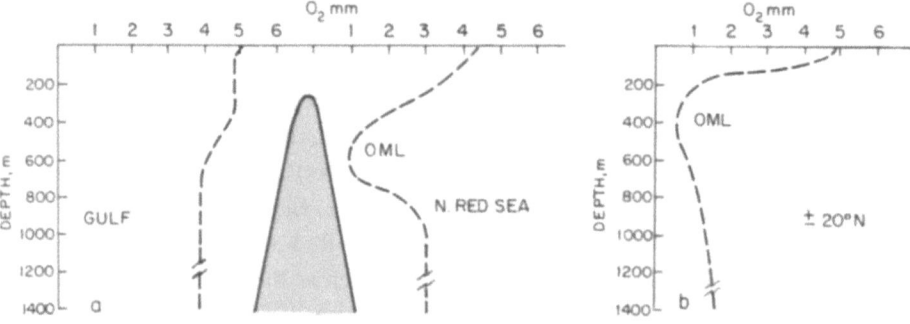

Fig. 4.4. Vertical distribution of dissolved oxygen in a) the Gulf of Aqaba and b) in the central Red Sea. OLM = Oxygen Minimum Layer (after Reiss and Hottinger, 1984 and Weikert, 1987).

It results that the Gulf of Aqaba has a very marked hydrographical specificity if compared to the main body of the Red Sea. It is first of all peculiar by the fact that all its water mass circulates for a few months in the year. Would it not be for its small size, the Gulf of Aqaba would be entitled on oceanographic grounds to be considered a separate sea of the world ocean.

The Gulf of Suez is the remnant of a geologically older and now quiescent rifting activity: it is most probably the oldest embryo of the Red Sea. It is a very shallow gulf, for most of its length only 20–30 m deep and plunging directly from a 90 m deep shelf into the deep Red Sea. The Gulf of Suez can be compared with the Persian Gulf of the Arabian Sea or the Adriatic Sea of the Mediterranean. Extending northwards along the desertic African and Sinai coasts, there is a sharp gradient of increasing salinity in the Gulf of Suez. Paraphrasing Morcos (1970) the increase of salinity is of 1.25 ppt (or even 1.50 ppt, Por, 1978) for one degree of latitude as compared to 0.25 ppt per one degree in the rest of the Red Sea. Values as high as 45 ppt were measured in the shallow waters of the Gulf of Suez shores.

Whether this high salinity is exclusively due to evaporation (over 300 cm/yr) or also due to the leaching out of old salt deposits, like in the Bitter Lakes of the Suez Canal, is not relevant to our subject.

Because of the shallow waters, temperature fluctuations in this gulf are very extreme and the waters are often extremely turbid because of suspended sediments. Low winter temperatures of 16°C are not exceptional in the open waters.

Finally, and most importantly: the Gulf of Suez became dry during the repeated stages of low eustatic levels and/or in the intermediate phases shrunk into the shape of a chain of "Bitter Lakes" (Por, 1978) with various degrees of metahaline to hypersaline salinities. The newly flooded salt swamps and lagoons were resettled from the Red Sea at each Glacial Termination. Unlike this, the large water mass of the Gulf of Aqaba maintained stable marine conditions during the Glacial low eustatic sea level. Its waters were probably cooler or saltier than at present but still hospitable to marine life (Reiss and Hottinger, 1984; see above, p. 29).

The Gulf of Suez had to be resettled each time by the coral species. Even today they are unable to build a real massive reef north of At Tur on the Sinai coast (Amittai and Raz 1969). In the Gulf of Aqaba, some coral species probably survived during the glacial phases and only the building of the coral reefs had to start each time anew; today there are coral reefs all along the coasts of this gulf. Mergner and Schumacher (1981) believe that given the hydrographic conditions of the deep Gulf of Aqaba and its "homeostatic" potential, coral reefs could grow in it even if its waters were to extend further inland and further north.

The pteropod *Creseis acicula* survived in the waters of the glacial Gulf of Aqaba (Almogi-Labin, 1982) accompanied by the coccolithophorid *Gephyrocapsa oceanica* (Winter, 1982). This indicates that besides these two calcareous plankters there must have been a fairly diverse life of other non-calcareous planktonic organisms.

Reiss and Hottinger (1984) also consider the possibility that during the last Glacial, the Gulf of Aqaba was more productive, and less oligotrophic than today. Luz et al.(1984) even speak of temporary anaerobic conditions on the bottom of the Gulf during the last glacial climax.

One has to think of the possibility that the Gulf of Aqaba, with a better supply of freshwater from the Sinai mountains, for some of which even periglacial conditions and temporary snow cover are accepted, could serve as a refugium for certain Red Sea fauna during the glacial crisis (see also below, p. 81). The duration of the seasonal homothermy must also have been longer during the Glacial heights.

4.4. Origin of biota and biotic provinciality

The modern fauna of the Red Sea is of exclusive Indopacific origin. For much time this seemed to be almost paradoxical since the low and narrow Isthmus of Suez, separating from the Mediterranean and covered by various waterbodies seems such a flimsy barrier. Moreover, accepting the existence of high Interglacial eustatic sea levels of tens of meters above the present level, it seemed evident that the Isthmus would have been repeatedly submerged. Very scarce proof for such a Pleistocenic connection could be found, however. Today, thinking in terms of Interglacial sea levels never much higher than the present ones, this riddle is solved (see above, pp. 22–23). Another less persistent view that Gulf of Aqaba transgressed into the the Dead Sea valley belongs to the same category of hypotheses, which for the time being can be safely shelved.

The modern man-made Suez Canal connection, which induced the spectacular phenomenon of the Lessepsian migration of Red Sea biota into the Mediterranean (see p. 55) also motivated discussions about an eventual "Anti-lessepsian" migration from the Mediterranean into the Red Sea (Por, 1978). If such influx indeed occurred, it was of a very limited and local impact. As shown by Por (op. cit.) only very euryhaline Mediterranean species crossed the Suez Canal in a southern direction and in general remained confined to the northern part of the Gulf of Suez or its lagoons. Such are the fishes *Dicentrarchus punctatus, Liza aurata, Sciaena aquilla, Umbrina cirrosa, Engraulis enchrassicolis* and *Hippocampus brevirostris*. In general these Mediterranean fishes survive in a few marginal habitats of the Gulf of Suez (Ormond and Edwards, 1987). In a recent paper dealing with the possibility of Anti-lessepsian migration in mollusks, Barash and Danin (1987) list very few convincing cases: *Cerastoderma glaucum* and *Pirenella conica* belong to the very euryhaline "isthmus fauna" (Por, 1978); *Arcularia gibbosula (=Nassa gibbosula)*, which is a Mediterranean endemic lives also in the Canal. All three species did not expand beyond the Gulf of Suez. Barash and Danin (op. cit.) also list 6 species of chitons, which, though not clearly Suez Canal migrants, could have be taken across on ship hulls. Other species are either known also from the African Atlantic coasts or are too widespread in the Indian Ocean to be recent faunal additions. Rather intriguing is the presence of the peculiar and quite obvious Mediterranean sea star

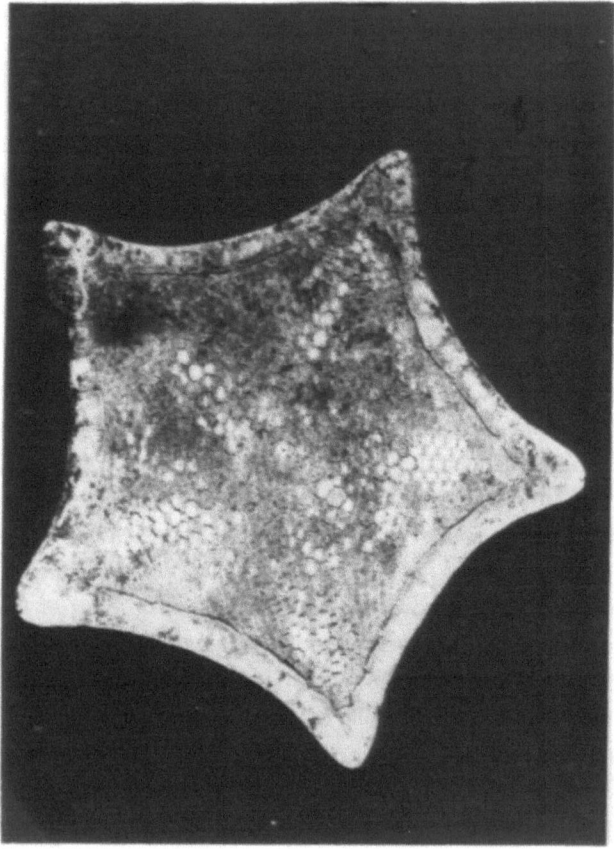

Fig. 4.5. The supposed Antilessepsian migrant *Sphaeriodiscus placenta* (from Fouda and Hellal, 1987).

Sphaeriodiscus placenta at El Ghardaqa (Fouda and Hellal, 1987) (Fig. 4.5). Further data on this interesting potential Anti-lessepsian migrant are very much needed.

One could expect that the Red Sea with its steep environmental gradients would present a clear-cut picture of faunal depauperation from south to north. This is, however, not unequivocal. For several years, the north and especially the Gulf of Aqaba was much better known than the Central and southern Red Sea. During the last decade, much information became available, especially for the Central part: therefore the danger of hasty conclusions based on unequal information is today much reduced. Yet, the southernmost Red Sea remains insufficiently known.

The situation as we see it seems to be rather complex: besides a gradual decrease in diversity from south to north, one finds also some "recovery" in the northern Red Sea and even "enclave" situations in the Gulf of Aqaba.

Compared with the Indian Ocean, the Red Sea as a whole is much poorer in species: but again within the Red Sea itself, the situation is less clear.

A clear trend of impoverishment is especially visible in the intertidal and in the mangal or mangrove communities. In the southern Red Sea there are three mangrove tree species: *Avicennia marina, Rhizophora mucronata* and *Bruguiera gymnorhiza* (Zahran, 1977). The last two species reached only into the Central Red Sea.

Fig. 4.6. Shurat el Arwashie – the largest Mangal area in southern Sinai (photo D. Darom).

Along the eastern coast of Sinai, where at 28° N (Ras Atantur, Por et al., 1977) the mangrove community reaches its northern limit, there are only shrubs of *Avicennia marina* left. In the Gulfs of Suez and most of the Gulf of Aqaba there are no mangroves anymore. This corresponds to a pattern known worldwide: it is related with the decreasing air temperatures and as a rule, a species of *Avicennia* is the last mangrove tree to disappear.

Por et al. (1977) and Por (1984a) characterized the Sinai mangal as a "metahaline – hard bottom mangal" contrasting with the "estuarine – soft bottom mangal", which is the prevalent type of mangal worldwide. The metahaline mangrove reaches its limits at about 50 ppt water salinity. The development of the mangrove in the Red Sea gulfs is limited by the very high salinity of the water in the intertidal sediments (Por, 1984), besides the limitation by low air temperature. Again, *Avicennia* is the most salinity-resistant genus of the mangrove trees (Figs. 4.6 and 4.7).

There is also a concomitant decrease in the diversity of the animal species associated with the mangrove: in Sinai only two species of fiddler crabs survive, namely *Uca tetragonon* and *U. inversa* (Por et al. op. cit.). A third species, namely *Uca urvillei* is reported from the Saudian coast by Price et al. (1987). In the southern Red Sea, the diversity increases even more. According to Jones (1984), *U. urvillei* has a lethal ambient temperature of 40°C and is shade-loving, whereas the two Sinai species are resistant to a higher temperature of 43°C and are less restricted to shade. Typical mangal species like the mudskipper *Periophthalmus*

Fig. 4.7. Mangal in southern Sinai (photo D. Darom).

and the mangrove snail *Cerithidea cingulata* do not reach the northern Red Sea either. The lack of *Rhizophora* with its stilt roots might also be limiting for the presence of the mudskippers.

A similar sharp gradient of impoverishment is found in the distribution of the sea grasses (Lipkin, 1977; Jones et al., 1987): Of the 10 species known from the open Red Sea, 5 or 6 species are still found at the entrance of the Gulf of Aqaba, while only 3 species, namely *Halophila stipulacea*, *H. ovalis* and *Halodule uninervis* grow at the northern end of this Gulf. Only *H.stipulacea* and *H. uninervis* reach the tip of the Gulf of Suez (Fig. 4.8).

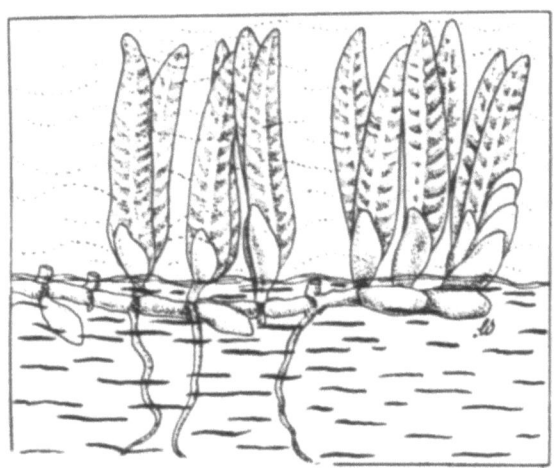

Fig. 4.8. The Seagrass *Halophila stipulacea*.

Primary production values show that the relatively eutrophic conditions in the Gulf of Aden and the southern Red Sea gradually turn oligotrophic as one advances northward. Weikert (1987) compares the values presented by various authors, using, however, different units of measurement. According to Reiss and Hottinger (1984) the productivity in the Gulf of Aqaba is five to ten times lower than in the southern Red Sea. Extreme oligotrophic values in this Gulf are comparable to those of the oceanic gyre centers (Sournia, 1977; Levanon-Spanier et al., 1979).

However, this trophic depletion seems to a certain extent to be compensated by peculiar hydrographic patterns in the Gulf of Aqaba. First of all, as mentioned above (p. 71), the Gulf waters present an annual period of three months of destratification with a consequent redistribution of the nutrients in the whole water column (Klinker et al., 1976). As a consequence, many phytoplanktonic algal taxa are blooming in the early spring months (Kimor and Goldansky, 1977), and the calanoid copepods as well show a spring peak (Almeida Prado-Por, 1983). The benthic algae in the northern Gulf of Aqaba and especially *Sargassum dentifolium* and *Cystoseira mirica* are also presenting peak biomasses during the nutrient enrichment of February-March (Mergner and Svoboda, 1977). The Gulf of Aqaba is a peculiar waterbody in which tropical biota are found under subtropical hydrographic conditions of strong, even if short, seasonality.

The Gulf of Aqaba does not have an oxygen minimum layer (OML) (see above). Weikert (1982, 1987) has described a very typical zooplankton which characterizes the OML in the open Red Sea. There is no such stratification in the Gulf of Aqaba and the calanoid zooplankton migrates from the deep layers to the surface without the barrier of a sharp oxycline (Almeida Prado-Por, 1980). According to Weikert (1987) some zooplankters are especially abundant in the northern Red Sea and the Gulf of Aqaba: such are the copepod *Rhincalanus nasutus* and the pteropod *Cresseis acicula*. According to Weikert (op. cit.) this is a situation opposed to the general pattern of impoverishment in the north. Possibly the Gulf of Aqaba harbors "old" species which survived there during the Glacials. Beckmann (1984) considers the calanoid *Haloptilus longicornis* to be typical for the northern Red Sea.

The Gulf has also a fairly diversified fish and zooplankton population which inhabits the mesopelagic zone (Aron and Goodyear, 1969; Almeida Prado-Por, 1980).

Halim (1984) distinguishes an "autochthonous" plankton assemblage in the northern Red Sea. Another assemblage, the one which is swept in from the Gulf of Aden by the winter monsoon flow, dies off in the transition to the Central Red Sea. Beckmann (1984) exemplifies this phenomenon by the massive collapse of the populations of the calanoid *Eucalanus crassus* in the area around 16–18° N. Casanova (1986) distinguishes a pattern which is exemplified by the arrow worms: *Pterosagitta draco* is an oceanic species which does not extend beyond the southern Red Sea; *Sagitta enflata* is found all over the sea and *S. pacifica* is characteristic for the northern, more saline part of the Red Sea.

An evident impoverishment in zooplankton occurs in the shallow and turbid Gulf of Suez (Halim, 1969). There are, however, also cases in which a specific neritic plankton develops there. Such is the case of the salp *Doliolum nationale* which is abundant in the Gulf of Suez and absent from the more oceanic Gulf of Aqaba (Godeaux, 1983).

It is still a puzzle why the deep and blue waters of the Gulf of Aqaba are inhabited by a full-fledged oceanic plankton, with practically no admixture of neritic taxa, up to the northernmost tip of the Gulf. Deep waters, steep shores, lack of run-off and extremely oligotrophic conditions are possible explanations, though at this stage they seem to be insufficient.

The level-bottom biomass of the shallow bottoms seems to be fairly high in the Red Sea. The values given by Murina (1971) are as follows: 37.3 g/m^2 at Bab el Mandeb; 26.5 g/m^2 in the southern Red Sea; 6.2 g/m^2 in the Central Red Sea and 21.6 g/m^2 in the northern Red Sea. Kisseleva (1971) considers that the meiobenthos of the Red Sea is more diversified than that of the Aegean Sea and some parts of the Adriatic and Black Seas. Thiel (1979) and Thiel and Weikert (1984) speak of low benthic biomasses but at the same time emphasize the presence of a specific fauna adapted to nutrient depletion and high bottom temperature. A relatively high diversity of the meiobenthos has been reported from the Gulf of Aqaba, where there is more dissolved oxygen on the bottom than in other parts of the Red Sea (Por and Lerner-Seggev, 1966; Por, 1967).

It is a remarkable fact that the interstitial meiobenthos of the northern Red Sea and of the Gulf of Aqaba is highly diversified and does not show qualitative depauperation, at least in normal, open-sea sites. Darom (1974) found a large variety of taxa in the interstitial of the Gulf, at salinities of over 44 ppt though biomasses were about 10 times lower than in "normal" interstitial environments. At El Ghardaqa, salinity values in the collecting sites were 45–46 ppt (Remane and Schultz, 1964). Nevertheless, Riedl (1966) found there about 65 species of Turbellaria, over 30 species of Nemertea, 7 of Gastrotricha and several of Kinorhyncha. Kisielewski (pers. comm.) found 9 taxa of Gastrotricha in one littoral sand sample near Eilat. Interstitial fauna is generally thought to be adapted to lower-than-sea-water salinities. In the Red Sea, where interstitial waters are even more saline than the open sea, the fauna does not seem to be qualitatively impoverished. It stands to reason that the interstitial animals which are probably very slowly colonizing, had to survive and adapt to high sediment water salinities during the eventful Pleistocenic history of the Red Sea.

There are no indications of impoverishment in the sublittoral macrobenthic fauna of the northern Red Sea and of the Gulf of Aqaba. The Gulf of Suez, with its highly saline, winter-cold and murky waters is again an exception. According to James and Pearse (1969) and Pearse (1983), from the 26 species of echinoderms living at El Ghardaqa near the entrance to this Gulf, only 17 species reach the northern end of it. Some extremely common Red Sea species like *Fromia ghardaqana, Echinothrix calamaris, Tripneustes gratilla, Heterocentrotus mamillatus, Ophiocoma pica* and *Actinopyga mauritiana* are missing. It is interesting to note that the species which survive in the Gulf of Suez, such as the sea urchins *Diadema setosum* and *Echinometra mathaei* have a short reproductive period restricted to the summer months: the other Red Sea populations of these species reproduce during a much longer season. Pearse (op. cit.) finds a similar restriction in the reproductive periods of the Suez populations of the abalone *Haliotis pustulatus* and of the chiton *Acanthopleura haddoni*.

In general, the reefs and the number of the species which build them do not show any marked impoverishment in the Gulf of Aqaba as compared to the main Red Sea (see below). The Gulf of Suez is again an exception: massive and actively growing reefs disappear north of At Tur and only small and isolated coral colonies are left. Scheer (1971) briefly described such a poorly developed reef on the African side, at Ras Shukheir. According to Scheer (1984) there are 36 genera of corals in the Gulf of Suez, compared with 56 genera in the Gulf of Aqaba. There are 7 genera (and subgenera) of corals of family Agaricidae in the Gulf of Aqaba and only one, namely *Pavona* in the Gulf of Suez.

4.5. The problem of the Red Sea endemism

The Red Sea has been considered by many authors as forming a separate biogeographic province within the Indopacific realm. Ekman(1967) mentioned the high percentage of endemic species among the Red Sea Decapoda, Crinoidea and Fishes. Briggs (1974) spoke of an overall level of 15% endemism in the fauna of the Red Sea.

This seems to contrast with the fact that the Red Sea was isolated from the Indian Ocean, or nearly so, during the last Glacial. For some authors it turned into a highly saline hypersaline lagoon (see above, p. 68). As it were, recolonisation with "normal" Indian Ocean fauna started only some 10,000 years ago. The problem of the Red Sea endemism has been discussed with ups and downs and considered at best as a difficult riddle to solve (Klausewitz, 1968; Por, 1975a). Lewinsohn (1969) even expressed scepticism about the presence of any endemics among the anomuran decapods of the Red Sea. Head (1987) speculates that the extraordinary burst of colonisation after the Glacial Termination must have provided a mechanism for accelerated evolutionary change.

Klausewitz (1983) correctly considered that the real area of distribution of the endemic Red Sea fishes includes often also the Gulf of Aden and sometimes even

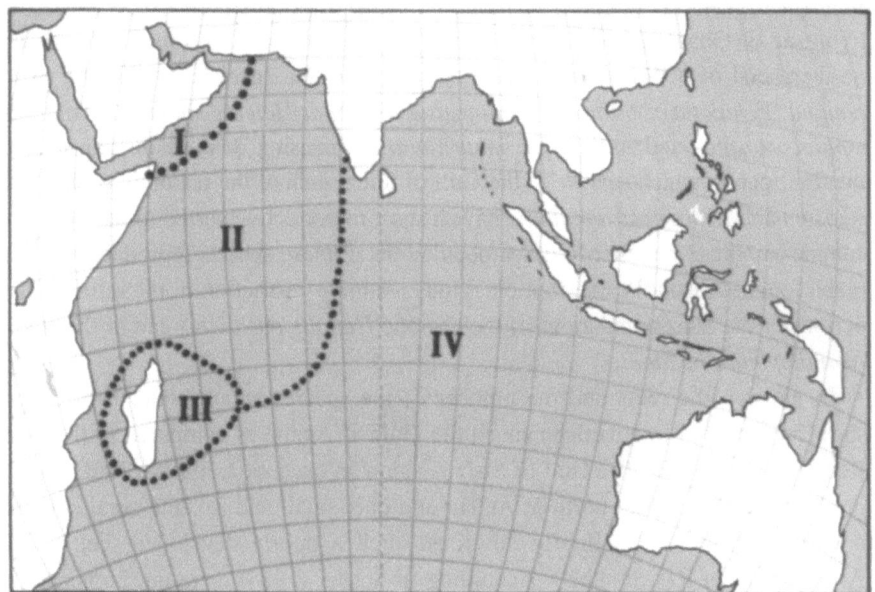

Fig. 4.9. Zoogeographic provinces in the western Indian Ocean. I – Red Sea-Persian Gulf; II – Western Indian Ocean; III – Malagassian; IV – Eastern Indian Ocean (after Klausewitz, 1983).

the Persian Gulf (Fig. 4.9). Thus, temporary extreme conditions in the Red Sea could act as a circumstance which enhanced speciation in the sense of Valentine's (1967) diversity pump, without endangering the survival of the species itself: repeated colonization of the Red Sea and "retreat" into the safety of the Gulf of Aden would have been a mechanism enhancing speciation (Por, 1975a).

Recent research has somewhat reduced the number of endemic species of the Red Sea, but confirmed the existence of others. For instance, the high number of endemic Crinoidea (70% according to Ekman, 1967) has not been confirmed recently. Recalculating the number and percentage of endemic shallow-water echinoderms in the Red Sea using Clark and Rowe (1976), Campbell (1987) calculates a proportion of 14% of endemic species. Most of the 27 endemic species are among the Holothuroidea: 15 out of a total of 64 species (i.e. 22%) and the lowest number of endemics is found among the Echinoidea. Campbell (op. cit.) considers that Ekman's high figure for the Crinoidea must be based on deep-sea species; but Thiel (1987), analyzing the deep-sea fauna of the Red Sea, states that Crinoidea are missing so far from the collections known to him.

Among the very rich fauna of Mollusca, even an approximate number of endemics is still uncertain. Mastaller (1987) mentions several endemic species of cowries (*Cypraea*) and of augers (*Terebra*) as well as *Strombus fasciatus*.

According to Mienis (pers. comm.), the cases of endemism among the Red Sea mollusks are as a rule at the subspecific level, though there are some problematic endemic species, too. Based on the knowledge of the better-known gastropod families Strombidae, Cypreidae and Conidae, a pattern similar to that of the fishes emerges: the areas of the typically Red Sea mollusks include also Gulf of Aden, Gulf of Oman and the Persian Gulf. The southern limit of this distribution pattern would be Cape Guardafui on the Somali coast. In other gastropod taxa, such as for instance *Rhinoclavis* (a new partial synonym of genus *Cerithium*), all 5 species of the Red Sea have a broad Indopacific distribution and are reported from all the Red Sea and both its gulfs (Barash, Danin and Yaron, 1984). It is interesting that only one of these widespread species, namely *Rh. kochii* penetrated into the Mediterranean through the Suez Canal.

Klausewitz (1983) took another step forward, distinguishing three stages in the process of "endemisation" of the Red Sea fishes: 1. subspecific endemics are of post-Glacial age; 2. endemics with an obvious sibling in the western Indian Ocean would be of an undetermined Interglacial age; and finally 3. species with no near relative in the Indian Ocean would have to be considered to be of Pliocenic age.

The same author also connects the percentage of endemism in the different families of fish with the larval biology of these families: it can be as high as 29% in the reef-haunting Pomacentridae and even 46% in the Chaetodontinae, but in the Muraenidae or Scorpaenidae which have pelagic larval stages, there are no endemic species in the Red Sea (Klausewitz, op. cit.). Also fishes which live in floating algal clumps, like *Paramonacanthus oblongus* can be passively swept into the Red Sea by the monsoon currents.

Ormond and Edwards (1987) consider that 17% of the fish species are endemic to the Red Sea. This proportion increases when one adds the species which are found also in the Arabian Sea and the Persian Gulf. Endemism is very high in Tripterygiidae (91%) and Pseudochromidae (90%). Levels of endemism are intermediate in Chaetodontidae, Blenniidae and Pomacentridae. Ormond and Edwards (op. cit.) further indicate that several endemic Pseudochromidae are confined to the northern Red Sea.

In general there are relatively few endemic species among the stony corals. According to Head (1987) only 5 species are likely candidates for the endemic status. Although this author considers the Red Sea to have been recolonized with corals only after the last deglaciation, he believes that genus *Stylophora* may be presently speciating in the Red Sea.

Among the algae there is a certain number of endemics. Two of these, *Turbinaria elatensis* and *Dichotrix eylathensis* seem to be confined to the Gulf of Aqaba (Walker, 1987). The first of these species is so conspicuous that an oversight of it in other parts of the Red Sea seems unlikely.

The existence of the Pliocenic "paleoendemics" of Klausewitz (1983) requires the permanence, even of restricted marine conditions in the post-Pliocenic Red Sea. Interesting in this respect is the case of the coral genus *Siderastrea* which according to Scheer (1984) was present in the Pliocene Mediterranean (!). Today the genus is represented in the tropical Atlantic, in the Red Sea and the western Indian Ocean. As shown by Por et al. (1977) *Siderastrea* is present in the metahaline mangrove lagoons of southern Sinai. It could be therefore a good example of a coral which survived "in situ" the Pleistocenic adversities.

Klausewitz (1983) also mentions as a likely Red Sea paleoendemic the fish *Zebrasoma xanthurum*, which resembles the Eocene *Naseus nuchalis* from the Monte Bolca slates in Italy. Another paleoendemic listed by this author is the shallow water goby *Lotilia graciliosa*.

The most likely candidate for the role of a shelter for marine species during the Glacial pessima is the Gulf of Aqaba, for reasons already mentioned above. Mastaller (1987) considers this gulf as a small marine branch which was isolated several times during the Pleistocene from the main basin and might have evolved its own faunal adaptations. As stated earlier by Por (1978), the Red Sea endemics have to be first of all sought among the high-salinity and low temperature adapted species of the northern Red Sea and the Gulf of Aqaba. However, this hypothesis is not yet based on sufficient information.

The whole issue of endemism in the Red Sea gained a new dimension with the recent data from the deep sea.

4.6. A warmwater deep-sea fauna

As mentioned before, minimum temperatures near the deep-sea bottoms of the Red Sea do not fall below 21°C. In a sense one might consider the deep Red Sea to be a model for the Cretacic deep oceans, before the dramatic Cenozoic temperature decrease.

Ever since the time of the "Pola" Expedition which discovered this unusual situation, the existence and the specificity of the deep-sea fauna of the Red Sea was a debated issue.

Soon it became clear that there is a very low-density fauna on the deep bottoms, much poorer than in a normal ocean. Along with the high temperatures, the trophic conditions are most probably near starvation. The basic oligotrophicity of the Red Sea, combined with the rapid decomposition of the organic morsels at the high water temperatures leads to a situation in which very little food reaches the bottom (Wishner, 1980). There is probably no deep-sea plankton in this sea (Weikert, 1987) and even the areas of the hot brines do not show increased biomasses (Wishner, op. cit.).

Still, the "Pola" material contained bivalves of genus *Cuspidaria* typical of deep bottoms elsewhere (Sturany, 1899) and species of this genus were later found also in the depth of the Gulf of Aqaba (Zalcman, unpublished thesis). Balss (1915, 1929) studied decapod Crustacea from the deep-sea catches of "Pola". He concluded that the deep-sea fauna of the Red Sea is essentially an infralittoral, eurybathic fauna of the Indian Ocean which extended its depth range under the special conditions of the Red Sea. Türkay (1986) revising the deep-sea decapods of the Red Sea reports several cases in which populations of Red Sea crabs live at much shallower depths in the Indian Ocean. However, the list of clear-cut examples is already reduced to three species: *Munida japonica*, *Paguristes calvus* and *P. incomitatus*. The other cases are sibling species or unclarified cases. Mastaller (1987) accepts the older data according to which the gastropods *Murex tribulus*, *Nassarius albescens* and *Cantharus fumosus*, shallow-water species in the Indo-Pacific have a deep-sea distribution in the Red Sea.

The deep-sea fauna of the Red Sea is probably a mixture of such "submergent" Indo-Pacific species along with endemic species. Por (1967), working on the bathyal harpacticoid copepods of the Gulf of Aqaba, found there genera like *Cervinia*, *Eurycletodes* and *Mesocletodes* common in the bathyal and abyssal faunas in the world ocean. But some typical abyssal genera like *Pontostratiotes* were not found; on the other hand the deep-water genera *Ophirion* and *Hypalocletodes* described from the Gulf of Aqaba have not yet been reported from elsewhere.

Mortensen (1939) described from the northern Red Sea and the Gulf of Aqaba peculiar deep-sea echinoid *Pericosmus akabanus*; another interesting echinoid of the Gulf is *Palaeostoma mirabile*; both species are related to Pliocenic fossils. The second species was later found as a very typical component of the deep bottoms of around 500 m in the Gulf of Aqaba (Por and Lerner-Seggev, 1966).

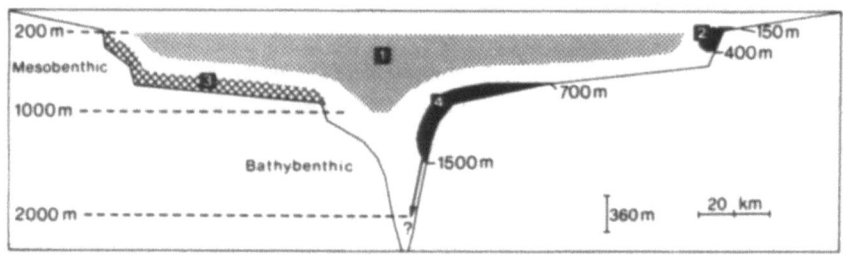

Fig. 4.10. Schematic distribution of categories of deep-sea fish in the Red Sea. 1 – Midwater fishes; 2 – Deep-bottom dwellers; 3 – Eurybathic bottom dwellers; 4 – Deep-sea inhabitants (from Klausewitz, 1986).

The last years saw an increase in our knowledge of the deep-sea fauna of the Red Sea. This is related to the work done by the "Meteor" and the Saudi-Sudanese project MESEDA, resulting in a series of publications by German oceanographers. Türkay (1986) reports several bathyal and abyssal decapods which appear to be endemics. The deepest record is that of *Solitariopagurus profundus* found between 1300–1995 m. Another new species is *Viaderiana meseda* found at depths of over 1000 m in the northern Red Sea, whereas its congeneric relatives in the Indopacific are shallow-water species. The case of *Achaeus erythraeus*, found in the Gulf of Aqaba by the "Pola" and described by Balss (1929) and now found as an extremely common species between 363–1203 m is similar. *Nematopagurus lewinsohni* is a bathyal species also found in the Gulf of Aqaba. Finally, *Nematopagurus helleri* described by Balss, and now found frequently in the Central Red Sea seems to be an exceptional case: it already shows morphological adaptations typical for a deep-sea species; also its nearest relative is the Atlanto-Mediterranean *N. longicornis*. This would eventually be an example of paleoendemism.

A personal communication by Türkay, quoted from Thiel (1987) mentions two endemic prawns *Haliporus steindachneri* and *Parapandalus adensameri*. Türkay (1986) considers that 30% of the deep-sea decapods are endemic species.

Klausewitz and his colleagues and Ben-Tuvia and his school have been publishing since 1980 a series of papers on the deep-sea fishes resulting from the MESEDA project in the Central Red Sea and from the Gulf of Aqaba respectively. Recently, Klausewitz (1986) reviewed the ichthyological knowledge accumulated, and presented an ecological hypothesis that explains the endemism of the deep-sea fishes of the Red Sea (Fig. 4.10). Accordingly in the category of the *midwater fishes* there are two cases of endemism, namely *Vinciguerra mabahiss* and *Astronesthes martensii*. The first of these species has been reported by Aron and Goodyear (1969) from the Gulf of Aqaba and is considered by Klausewitz to be typical for the northern Red Sea. In the second category, that of the *deep-bottom dwellers* (i.e. fishes living between 150–400 m) there are many Indian Ocean species which probably can freely pass over the shallow Bab El Mandab sill. One of these, *Chaetodon jaykari* has also passed the sill of Tiran and has been reported

Fig. 4.11. Coral reef in the Gulf of Aqaba (photo D. Darom).

from the Gulf of Aqaba (Klausewitz and Fricke, 1985). Neither are there any endemics among the third category, namely of the *eurybathic bottom dwellers*. Two deep-sea sharks included in this category have also been found in the Gulf of Aqaba: *Mustellus mosis* (Baranes, 1982) and *Iago omanensis* (Baranes and Ben Tuvia, 1979). Finally, the fourth group, the *deep-sea inhabitants* which live below 700 m, contains only endemic species. Since this group is still under intensive investigation, it is difficult to say what is its proportional weight among the totality of the deep-sea ichthyofauna, and in which part of the Red Sea it is mainly found. For our discussion here it is sufficient to assume that the deep waters of the Gulf of Aqaba are also likely to harbor such endemics.

Data for other taxa are still accumulating. Andrea (1981) described three new deep-sea amphipods, but their absence in the Indian Ocean has still to be established.

4.7. The paradox of the Gulf of Aqaba reefs

Flourishing and species-rich coral reefs are found up to the northern end of the Gulf of Aqaba at 29°30' N. They are the northernmost Indopacific coral reefs. This is due without doubt to the fact that the deep Gulf waters maintain a high temperature of no less than 21°C, in a climatic zone where air temperatures in the winter sometimes reach freezing point (Fig. 4.11).

The well-investigated reefs of the Gulf contain 56 genera and subgenera and over 130 species of hermatypic Scleractinia. The northern Red Sea has a more reduced inventory of 51 genera; the Central Red Sea an optimum of 60 genera and the southern Red Sea only 34 genera. For the Arabian Sea, Scheer (op. cit.) indicates 12 genera only. Even if these areas will be better investigated and more species found, it is certain that there is no impoverishment of coral diversity in the extreme north of the Red Sea (Fig. 4.12).

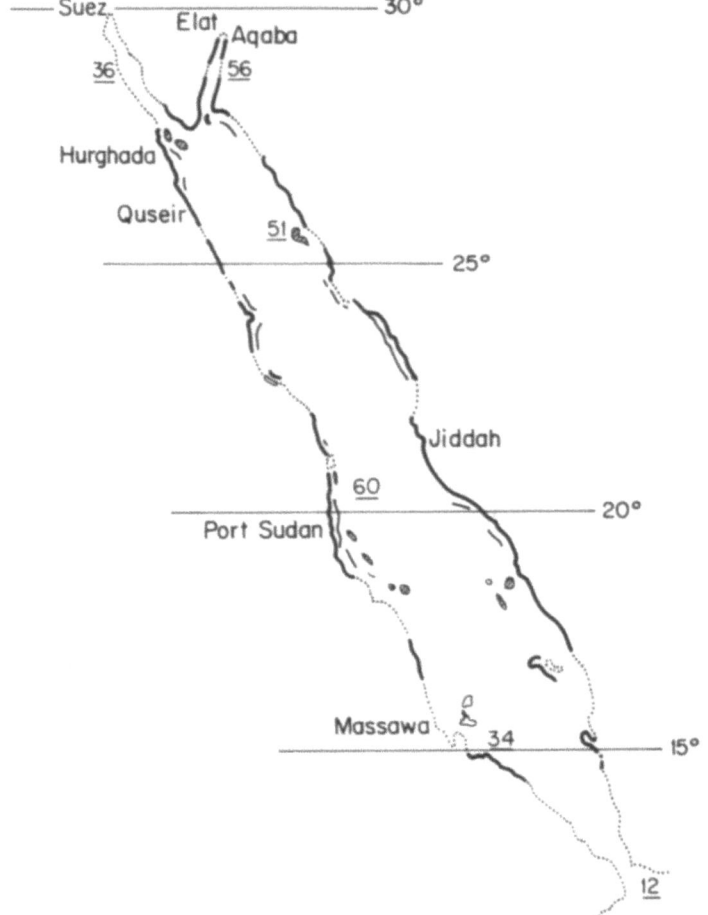

Fig. 4.12. Major coral reef coasts and coral-surrounded islands in the Red Sea. Diversity of coral reef genera is indicated (after Head, 1987 and Scheer, 1984).

In the narrow and steep-walled Gulf, reefs are only of the "fringing reef" type: with a maximum width of a few tens of meters and sometimes barely of 3–4 meters. The Central and southern Red Sea present belts of fringing reefs of several kilometers (for example the Towartit reef in Sudan, Head, 1987), many coral-fringed islands and even atoll-like formations. Yet, the dimension of the reefs does not seem to influence the diversity of the species.

The wealth in coral species in the northernmost Gulf of Aqaba, despite of its location at nearly 30° N, is among the highest in the world (Scheer and Pillai, 1983; Scheer, 1984). The diversity of scleractinian coral species in the Gulf of Aqaba reef, with an average of 15 species on a 10 m transect, is considered by Loya (pers. comm.) to be the highest for the coral reefs in the world.

Schumacher and Mergner (1985) made a comparison of the coral diversity indices per 25 sq meters of reef: the reef at Aqaba numbers 98 species whereas the Sanganeb reef in Sudan has only 62 species (Fig. 4.13). Further south, the predominance of reduced numbers of species over larger units of surface becomes even more marked. In the view of Schumacher and Mergner, this is due to the fact that the reef community in Aqaba does not achieve a real successional climax, which as a rule is accompanied by a somewhat reduced species diversity. Fishelson (1973), Loya (1976) and Benayahu and Loya (1977) discussed the impact of irregular catastrophies in the Gulf, like the extremely low tides of September 1970, which during four days exposed and killed the reef-flat population near Eilat. Such successional set-backs would maintain a high species diversity and retard the expansion of the "climax" species. The authors consider also some other factors, like concurrence with soft-corals and algae and grazing by sea urchins, as eventual causes for successional set-backs. Fishelson (1977), in a general discussion of the subject, theorizes that environmental disturbances are propitious for higher species diversity, if they remain within certain limits of amplitude and frequency.

Another circumstance which might explain the existence of an intricate niche partitioning on the narrow reef ledges of the Gulf of Aqaba has been put in evidence by Shlesinger and Loya (1985). Accordingly, the reef species of the Gulf have a reproductive strategy radically different from those of the Great Barrier Reef. In the Gulf, each species has a spawning period of its own, limited to few days and not overlapping, whereas on the Barrier Reef many species broadcast their larvae synchronously and for extended periods. In the Gulf, the spawning of the dominant species of corals is coordinated, whereas on the Great Barrier Reef, over 100 species might spawn during the same night.

This reproductive timing might be one of the ways in which a high diversity per unit surface is maintained. Furthermore, studies by Loya and his school have shown that reproduction of the stone corals in the Gulf of Aqaba is highly seasonal and restricted to the four summer months. During the winter circulation, which is so characteristic for this northern Gulf (see above), algal cover develops on the reef. The dying-off of the algae towards the summer, provides plenty of empty space for the settlement of the young coral colonies of the summer.

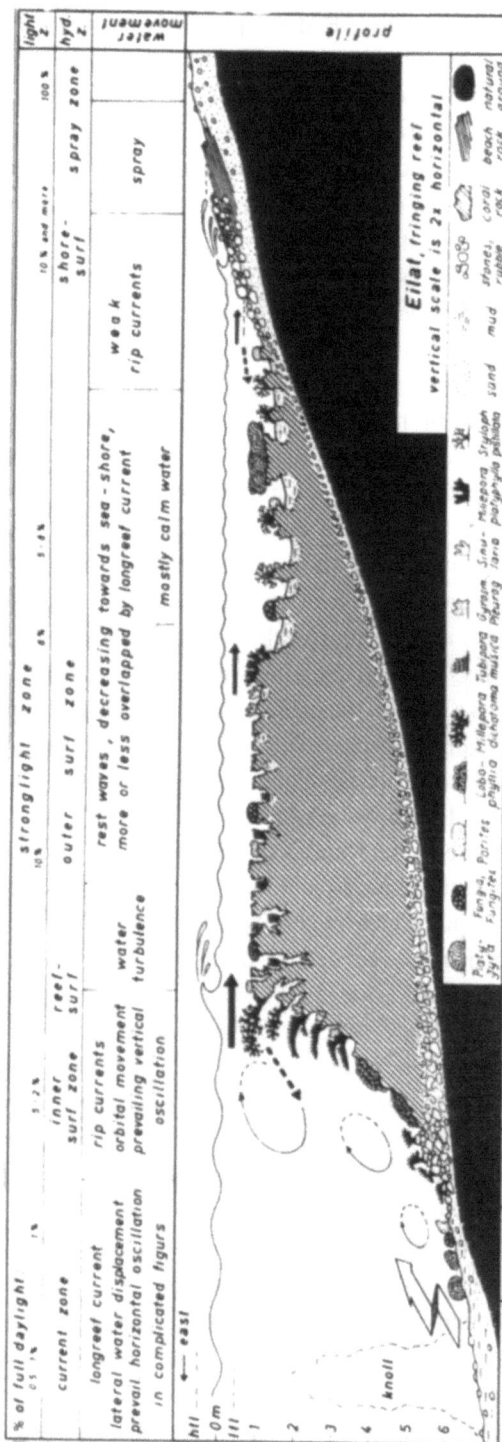

Fig. 4.13. Section through the coral reef of Eilat (from Mergner, 1971).

In the clear waters of the Gulf of Aqaba, hermatypic coral species can grow to considerable depths. Loya (1972) actually found that the number of species increases from the surface to 40 m depth. Whereas off Port Sudan, below 50 meters there are only 6 coral species (Kuhlmann, 1983), the situation is different in the Gulf: Investigations using a reasarch submersible (Fricke and Hottinger, 1983; Fricke and Schumacher, 1983) recorded no less than 47 species of corals below 40 meters. Such a diversity of deep-water hermatypic corals is equalled only by the deep forereef of Bikini attol. The Genus *Leptoseris* with several species is typical for the deep reef of Eilat and one of the species reaches its depth limit only at 145 m. This depth tolerance of hermatypic corals is reportedly exceeded only on Easter Island. The reason for the great bathymetric range of the hermatypic zooxanthellate corals near Eilat is to be found in the fact that the irradiance level of 1% is reached only around 100 m depth.

To explain the presence of a highly diverse reef in the enclave found in the northernmost corner of the Indopacific tropics, we feel that historical considerations have to be called upon. According to Braithwaite (1982) the recent reefs of the Sudanese Red Sea are only 5–6000 years old and 10 to 20 m thick. We believe that the reefs of the Gulf of Aqaba have a similar age. Fricke and Schumacher (1983) recognized four submarine terraces near Eilat: at 20–25, 40–45, 60 and 90–105 m depth. The last one would correspond to the low eustatic stand of the Last Glaciation and the terrace at –60 m is considered to be the basis of the present reef. The elevated fossil coral terraces on the Sinai side of the Gulf have been dated between 306,000 and 81,000 years old (Reiss and Hottinger, 1984). On the Island of Tiran there are at least 11 terraces up to an elevation of 320 m (Schick, 1958). Indeed, the high elevation of the terraces is a result of active tectonism.

If one accepts that during each Glacial the high salinities in the semi-isolated Red Sea exterminated the fauna of corals, then one has to admit an at least ten-fold recolonisation and reconstruction of rich coral reefs in the Gulf of Aqaba, at a distance of almost 2000 km from the Bab el Mandab straits. Furthermore recent studies about the larval biology of scleractinians show the prevalence of short-lived planulae with limited capacity of dispersal. Though one may count on several thousands of years for the successful re-establishment of coral reefs in the Gulf of Aqaba, (probably some 2000–3000 years), a stepwise advance of the larval propagulae from the Indian Ocean must have been at best a very difficult process.

The Gulf of Suez could be used as an actualistic model: it presents hydrographic conditions which are fairly close to those calculated for the last Glacial of the Gulf of Aqaba (see above). Even as far as the Port of Suez, a limited diversity of coral species survive in small colonies. They do not form the massive framework of a reef but rather a "coral community" (Wainwright, 1965) adventive on fossil reefs. Species of *Stylophora* and of *Siderastrea* are known to support higher than open Red Sea salinities and temperatures much below 20°C (Por, 1972; Por et al., 1977) (Fig. 4.14). Therefore we advance the hypothesis that isolated colonies of scleractinians could have survived in marginal, less restrictive environments, again and

Fig. 4.14. The Scleractinian coral *Stylophora pistilata* (photo D. Darom).

again serving as "in situ" nuclei for the reconstruction of the coral reefs after each Termination event.

In the case of the eastern Mediterranean we saw (above) that considerable gradients of impoverishment exist, as compared with the western Mediterranean. The total absence of the Gorgonaria is relevant to the biogeographic riddle discussed here. The poor fauna of the Levant basin of the eastern Mediterranean is without doubt a result of the fluctuations resulting from the last Glacial cycle and much less of present restricting circumstances. The probable difference from the conditions in the Gulf of Aqaba is the fact that a Postglacial sapropelic event severely depleted the fauna, in addition to the Glacial itself. The short time and long distance are still slowing down the resettlement of the Levantine waters. If resettlement had had to start from the more distant Atlantic Ocean, through the Gibraltar Straits, the present faunal poverty would be even more apalling.

In the Gulf of Aqaba there was most probably much less need, if at all, for a resettlement from the distant Indian Ocean, after the last Glaciation. If this were not so, the short time of 10,000 years and the long linear distance would have resulted in a biogeographic signature at least as strong as the one present in the Levant basin.

5. Halmyric Environments

5.1. Halmyrology – hydrobiology of waters with changing salinities

In most areas of the world ocean, lagoons, nearshore ponds and estuaries are predominantly brackish, i.e. with variable low salinities. In the Levant and along the whole Red Sea, such littoral waterbodies almost always have variable and higher-than-sea water salinities. A wider term is needed for the waters of changing and deviating salinities, irrespective of the sense of this deviation. We propose an old term, used many years ago by French hydrobiologists, namely "Halmyrology" (after "Halmyris", the classical Greek name of a complex of saline and brackish lagoons south of the Delta of the Danube).

The halmyric waters of the Levant have the tendency to become more saline than the adjacent sea. The diversity of these environments made it necessary to propose classificatory units never used before. The metahaline range of salinities, seen as separate from the higher salinity ranges of the hypersaline waters (Hedgpeth, 1957; Por, 1972) (see above p. 68) is a definiton which gained broad acceptance (Kiener, 1978; Plaziat, 1982; Gavish et al., 1985; Jones et al., 1987). Another proposal, to separate hypersaline waters into alpha-, beta- and gamma-hypersaline waters (Por, 1981) is not yet used very much (but see Gerdes et al., 1985). For our biogeographical discussion, however, extreme hypersaline environments are of little relevance: they are inhabited by a very low diversity of species with a cosmopolitan distribution. Recent data, however, tend to indicate that we are dealing there with cosmopolitic "species groups"; therefore in the future we might have to change our views on this subject, too.

5.2. Anchialine environments of the Red Sea

The subfossil Flandrian high reef platforms along the shores of Sinai are the "terra typica" of this recently recognized type of littoral waterbodies, the Anchialine pools (Holthuis, 1973; Por, 1985). After the discovery of the first of such pools in Sinai, namely the Solar Lake in the surroundings of Elat (Por, 1967), other anchialine water bodies followed suit. In general anchialine pools are nearshore basins which have only a subterranean contact with the sea. The seawater enters through the crevices of the coral platform and feeds the pool surfacing in the form of "springs" on the seaward shore of the pool or at greater depths. The physio-

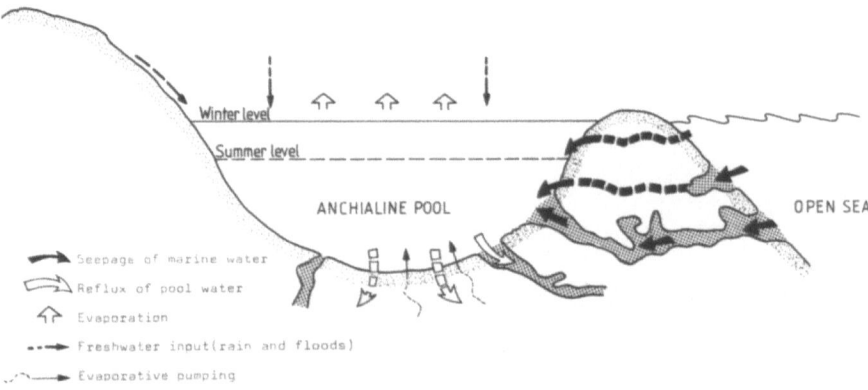

Fig. 5.1. Schematic representation of an anchialine Basin (from Por, 1985).

Fig. 5.2. The lagoon of El Kura on the Sinai coast, an anchialine pool in "statu nascendi". A slight further uplift of Sinai or a sea level fall will turn the lagoon into an anchialine pool (photo D. Darom).

graphy of each pool and its distance from the sea define the salinity and the dimension of its seasonal fluctuation (Por and Dor, 1975; Por, 1985). Since the freshwater inflow or run-off is extremely limited in our desertic area, all anchialine pools have higher than seawater salinities. Their maximal salinity is reached in the season of the highest evaporation, which is also the season of the lowest sea level. It may be important to mention that annual evaporation is above 3,000 mm and the critical months are September-October (Fig. 5.1).

Fig. 5.3. Calliasmata pholidota, a typical red-coloured anchialine shrimp (original).

Common to all anchialine pools is the long-term stability of their particular saline regimes since this is in function of their geomorphological setting. This is undergoing only long-term secular changes (Fig. 5.2). The result is the development of a high biotic diversity which is rather uncharacteristic for waters with higher-than-sea water salinities.

The fact that some of the anchialine pools may exhibit a chemical stratification and as a consequence a considerable adiabatic solar heating of the hypolimnion, like in the above-mentioned Solar Lake, is not important for our discussion here: the hot hypolimnion is also anaerobic and azoic. Por (1985) presented a formula for the calculation of the salinity regimes in the anchialine pools and supplied a first classification of them.

From the biogeographic point of view, the most interesting are the deep and shaded cracks in the reef, where salinity does not deviate markedly from that of the open sea. Several "cracks" have been investigated on Ras Muhammad, the southern tip of the Sinai Peninsula (Por and Tsurnamal, 1973).

These cracks contain an interesting faunal assemblage of red-coloured shrimps (Holthuis 1963; 1973). Among them *Antecaridina lauensis* is an interesting species: it is found also in anchialine reef cracks in the southern Red Sea, in the Indian Ocean and as far as Hawaii; it has never been reported from open marine waters. Other shrimp found in the Sinai cracks are *Calliasmata pholidota* (Fig. 5.3), a genus described from there but subsequently found also in anchialine environments of Hawaii, and the new species *Periclimenes pholeter* (Holthuis, 1973). Other peculiar species known till now only from the Ras Muhammad cracks are the goby *Cabillus anchialinae* (Klausewitz, 1975) and the keyhole limpets *Diodora*

Fig. 5.4. The anchialine pool of Di Zahav (from Por and Dor, 1975).

yaroni and *D. yaroni isaaci* (Christiaens, 1987). The copepods are represented by an undescribed species of the benthic calanoid genus *Pseudocyclops*.

The Ras Muhammad cracks are the first member of a sequence of anchialine pools with increasing salinities. Already the pool of Dahab (Di Zahav) is in the metahaline range with a seasonal salinity fluctuation between 45 and 60 ppt (Por and Dor, 1975) (Fig. 5.4). This anchialine pool of the Sinai shore of the Gulf of Aqaba is still inhabited by a restricted, but quantitatively luxuriant marine fauna, mainly growing on an impressive serpulid reef built by *Vermiliopsis pygidilis*.

Interesting for our discussion is the fact that the only coelenterate which lives in the Dahab pool is the upside-down medusa *Cassiopeia andromeda*. The place of the corals is occupied besides the serpulid worms, by a rich fauna of sponges and the ascidian *Ecteinascidia*. The meiobenthos of this metahaline waterbody is very rich. Among them the benthic calanoid copepod *Ridgewaya typica* (Por, 1979) stands out: this type of calanoids, freely gliding above the substrate and within the dark or half-dark cracks is typical to similar anchialine and marine cave environments in the Caribbean.

Another anchialine pool of the Sinai coast, the above-mentioned Solar Lake. By the classification proposed by (Por, 1985), this is an alpha-hypersaline waterbody. Literature about this less than 10,000 square meter-sized pool runs into more than hundred titles (Fig. 5.5).

The 6 m deep pool exhibits an extremely marked saline stratification for most of

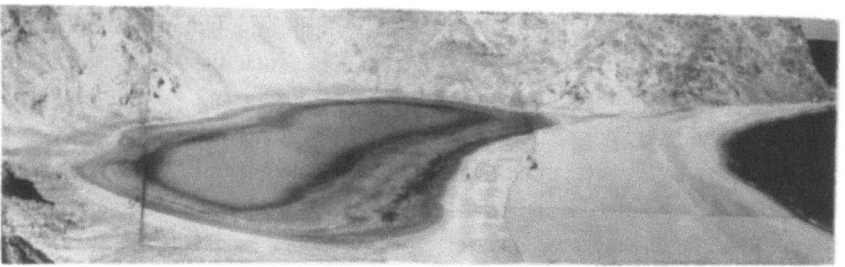

Fig. 5.5. General view of the Solar lake on the shores of Sinai (original).

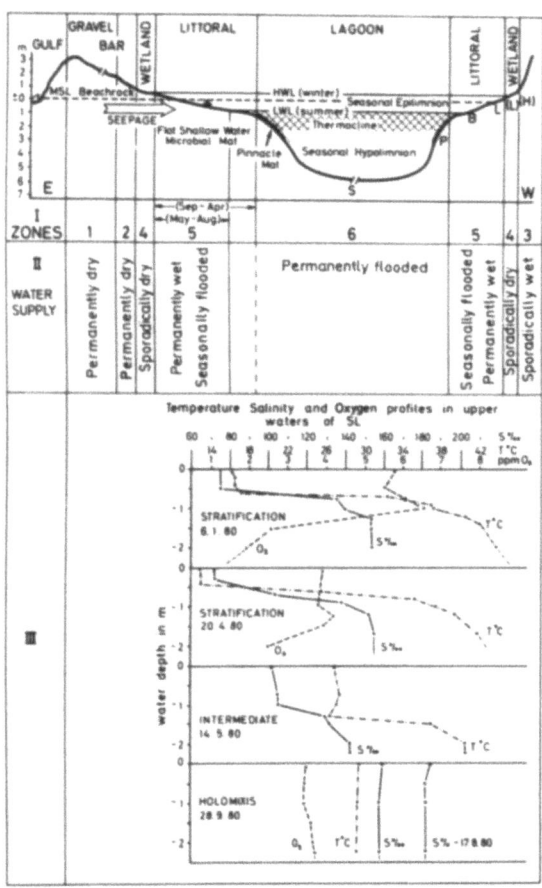

Fig. 5.6. The hydrography of Solar lake in Sinai (after Gerdes et al., 1985).

the year. The salinity in the epilimnion fluctuates from a minimum of 70 ppt during summer stratification, to a maximum of 120 ppt during the winter destratification (Fig. 5.6). Essential for life in the pond is the fact that the saline-water snail *Pirenella conica* cannot withstand a salinity over 100 ppt and that the bluegreen algae are consequently not restrained by grazing. The rich algal mats of bluegreens are inhabited by a very diversified fauna of aquatic coleopterans (Zalcman and Por, 1975; Por, 1975c; Gerdes et al., 1985). There are 11 species of coleopterans and their adult and larval stages occupy very specific niches each. Besides them, there are a few other hypersaline species. With the exception of the harpacticoid copepod *Robertsonia salsa* there are no marine species present. There is a local population of a parthenogenetic species of the brine shrimp *Artemia*, the only inhabitant of the open water pelagial.

There were very interesting attempts to calculate the radioisotopic age of the different evolutionary phases of Solar Lake (Krumbein and Cohen, 1974; Schidlowski et al., 1985). Accordingly, before 3500–4000 B.P. the area was a lagoon open to the sea and inhabited by *Pirenella*; after the separation from the sea the shallow lagoon started to accumulate the mats of bluegreen algae; around 2000 B.P., owing to a tectonic event, Solar Lake became a deep lake, with no bluegreen mat formation on the bottom.

A much larger but shallow anchialine waterbody is the Gavish Sabkha in southern Sinai, which served as a central object for the volume edited by Friedman and Krumbein (1978). This waterbody does not have a vertical stratification like Solar Lake, but in the horizontal sense exhibits a salinity gradient which increases from 50 ppt near the "springs" to 300 ppt in the more distant recesses of the shallow waterbody. Ehrlich and Dor (1985) could establish a very clear distributional gradient of the algal vegetation: among other facts they showed that the metahaline-hypersaline limit situated at 70 ppt is also upper limit for the green algae and most of the diatoms. Above this limit starts the prevalence of the bluegreen algae. The fauna of the Gavish Sabkha also harbours 10 species of coleopterans (Gerdes et al., 1985). However, because of the extreme shallowness, and eventually the limited area of the alpha- and beta-hypersaline zones, there is no local population of *Artemia* in this basin. Gavish et al. (1985) consider that the Sabkha came into being after a tectonic uplift some 2000–4000 years ago, which stranded the fringing reef. Round-shaped gaps, as seen in the modern reef, could have formed the anchialine basins of the subfossil stranded reefs (Fig. 5.2). In our opinion, rather than a tectonic uplifting, one should consider a marine regression from the higher "Flandrian" sea-level (see also Friedman, 1985).

Another very shallow anchialine basin is found on Ras Muhammad (Friedman et al., 1985): metahaline salinities exist there only near the springs. The majority of the basin is hypersaline and is seasonally dry. Without indicating salinity or biological data, Purser (1985) reports a hypersaline lagoon near Ras Gharib, on the African side of the Gulf of Suez.

Por (1975c) commented on the interesting fact that the hypersaline waterbodies of Sinai present a high diversity of aquatic and amphibious Coleoptera while the fauna of Diptera is very reduced and Hemiptera are missing altogether.

5.3. Metahaline lagoons of the Red Sea

Unlike anchialine waterbodies, the open-contact lagoons present smooth gradients from sea-water salinity to metahaline and even hyperhaline salinities. These lagoons are open to the influx of larvae and fishes from the open sea and are characterized therefore by banks of *Crassostrea* and *Mytilus*, and by schools of Mugilidae and of *Dicentrarchus*.

The shores of the Red Sea are dotted with countless small lagoons and drown wadi's. These are known under the local names of "Sharm", "Marsa" and "Ghor". They are too small to have individual faunal signatures and are as a rule contain the typical metahaline fauna of the Red Sea. Some shallow lagoons turn into salinas, called "Sabkha's".

Little is known about these inshore waterbodies of the Red Sea. As a rule, salinities of up to 47–48 ppt are encountered. Such is the case with the little lagoon on Faraun Island and El-Kura in the Gulf of Aqaba and of Ghor Blayim and Ras Matarma lagoon on the Gulf of Suez coast (Por et al., 1972).

More lagoons on the Suez side of Sinai have been investigated only by geologists: such are the lagoons of Ras Kanisa, Et Tur and Ras Lahata (Sneh and Friedman, 1985); unfortunately there are no salinity or biological data from these waterbodies. The lagoon of El Qardud on the Gulf of Aqaba side (Fig. 5.2) presents metahaline salinity values and a restricted marine fauna(Por, 1985), but unfortunately could not be studied in more detail. As mentioned already (pp. 74–76) the mangrove-bearing lagoons ofthe Nabq area in south Sinai present salinities of up to 48 ppt and extremely low winter temperatures of 10–13°C (Por et al., 1977). The presence of the scleractinians *Stylophora* and *Siderastrea* and of different alcyonarians and gorgonarians in these mangrove lagoons, together with echinoids like *Echinometra mathaei* and *Tripneustes gratilla* and a wealth of marine mollusks must be emphasized.

There are few data on other metahaline lagoons of the Red Sea. Meshal (1987) describes a lagoon from the eastern Red Sea shore, with salinities fluctuating in the annual range of 51 ppt to 113 ppt. Elsayed et al. (1985) studied the lagoon of Guemsah, on the African side of the Gulf of Suez with salinities ranging from seawater value to some 60 ppt. However, there is little biological information about these lagoons.

The metahaline coastal lagoons of the Red Sea deserve a more intensive investigation, since they are the key to the understanding of the Glacial Red Sea environments.

5.4. The Bitter Lake of the Isthmus of Suez

There is historical evidence from the writings of Strabo that during the Flandrian high sea-levels the Gulf of Suez was in contact with "Pikre Limne", the present Bitter Lake of the Isthmus of Suez. It was at that time the most important halmyric waterbody of the Gulf of Suez, an oversized "Ghor". The less than 20 km broad strip of land, with a maximum elevation of 2m, which separates the Lake from the Gulf of Suez today, must have been flooded time and again even before the construction of the Lessepsian Suez Canal (Por, 1978) (Fig. 5.7).

It is more adequate to speak of one Bitter Lake, rather than of artificial subdivisions into a Great and a Small Bitter Lake. The whole basin, filled by sea water through the Suez Canal, is today 36 km long, and has a maximum width of 13 km. According to Thorson (1971) it contains 85% of the waters of the Suez Canal system. Before the building of the Suez Canal, the Bitter Lake basin was a "sabkha", a salina, 7 km long and 2 km broad.

Lessepsian migration has been already discussed above (p. 55) and implicitly the role of the Canal and of the Bitter Lake as a pathway; we shall mention here the Bitter Lake as a habitat. In 1869, when the Lake basin was filled with sea-water, it started its new cycle of existence as a halmyric water body with a very poor autochthonous "isthmic" aquatic fauna (Por, 1971b). During the decade of 1880 the seawater which flooded the basin leached out the old salina to a salinity level at which already marine migratory species could penetrate into the Lake ("Hyperhaline phase"). When salinity fell below the 70 ppt mark, probably before

Fig. 5.7. Northern Suez Canal (aerial photo).

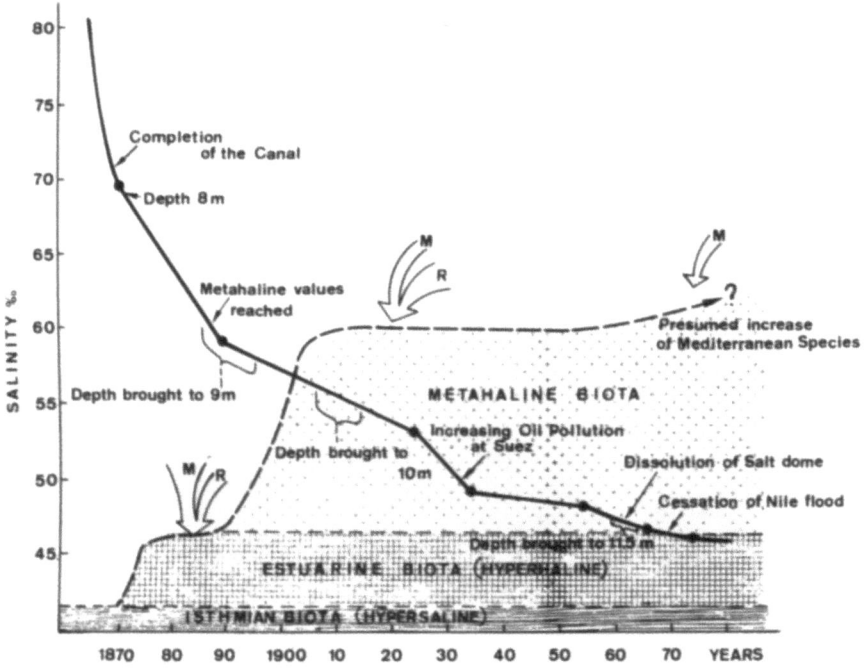

Fig. 5.8. Historical scheme of the main events in the process of settlement of the Suez Canal (from Por, 1978).

the turn of the century, the Bitter Lake became a metahaline lagoon of the distant Gulf of Suez (Fig. 5.8).

One can hypothesize that the Bitter Lake is nothing but the terminal member of a series of "Bitter Lakes" which became isolated or semi-isolated from the Gulf of Suez during consecutive stages of the retreat of the Glacial sea level (see p. 29). The inhabitants of the present, "youngest" Bitter Lake would be the usual pioneering metahaline biota of the Red Sea.

There are no natural rocky bottoms in the Bitter Lake and the waters of this large and relatively shallow, 15 m deep lake are turbid. Therefore little of the metahaline hard-bottom fauna entered the Lake. A stock-taking of the Bitter Lakes fauna is presented by Por (1978). In more recent years only the data by Aleem (1984) have been added.

This author considers the algal population in the Bitter Lake to be a typical Red Sea flora, without any influence from the Mediterranean. Among the new species in the Canal system, he lists the seagrasses *Thalassia hemprichi* and *Halophila ovalis*. Most interesting is the discovery by Aleem of small isolated colonies of stony corals, especially in the southern part of the Lake (Little Bitter Lake), despite a salinity of 49 ppt prevailing there. Before this report, only colonies of the

Fig. 5.9. The Nile Delta and the Suez Canal (satellite photo). The Bitter Lake is seen in the upper left half of the picture.

gorgonarian *Acabaria* were known from the Canal. Unfortunately, Aleem does not give the genus or species of these corals.

El-Sharkavy and Sharaf el Din (1983) confirmed the hypothesis expressed by Miller and Munns (1974) and by Por (1978) to the effect that the Bitter Lakes will always remain in the metahaline range. Because of the high evaporation of the waters of this shallow lake, salinities of over 46 ppt are still the rule.

The little we know about Lake Timsah, the other lake of the Suez Isthmus, will be mentioned in the following chapter.

5.5. Halmyric lagoons of the Nile Delta

There are today 5 delta lagoons, counting from the east: Bardawil lagoon; Lake Manzalah; Lake Burullus; Lake Edku and Lake Mariut. All these halmyric waters are connected with active or old Nile delta branches (Fig. 5.9). Bardawil represents the terminal lagoon of the old Pelusiac branch (Sneh and Weissbrod, 1973), Lake Burullus of the old Tanitic branch. Lake Timsah, a further Nile lagoon, is situated on the Isthmus of Suez at the outlet of another ancient deltaic branch, Wadi Tumilat.

Being supplied with Nile waters and separated from the sea by bars of sandy deltaic sediments, the Nile lagoons suffered extensive level and salinity fluctuations even in historical times. Recent engineering activity induced further changes: lake Manzalah was separated from the Bardawil Lagoon by the banks of the Suez Canal; the Ballah lagoon which connected Manzalah and Timsah, was drained. Lake Mariut, which dried out between the twelfth and eighteenth century, was artificially filled with water (and part of it transformed into the "Hydrodrome" of Alexandria). Following the building of the Aswan Dam, most of the freshwater reaching the four western lagoons was replaced by agricultural drainage (see above). Changes are also occurring in the physiography of the lagoons because of the radical decrease in the depositional activity of the Nile following the building of the dam.

The best investigated of the lagoons is the most saline Bardawil lagoon of northern Sinai (the Sirbonic lagoon of the classical and medieval authors) (Fig. 5.10). Por and Ben Tuvia (1982) summarize the knowledge about this lagoon, contained in more than 40 publications. The Bardawil occupies a surface of 650 square kilometers (maximum length of 90 km and maximum width of 22 km). The

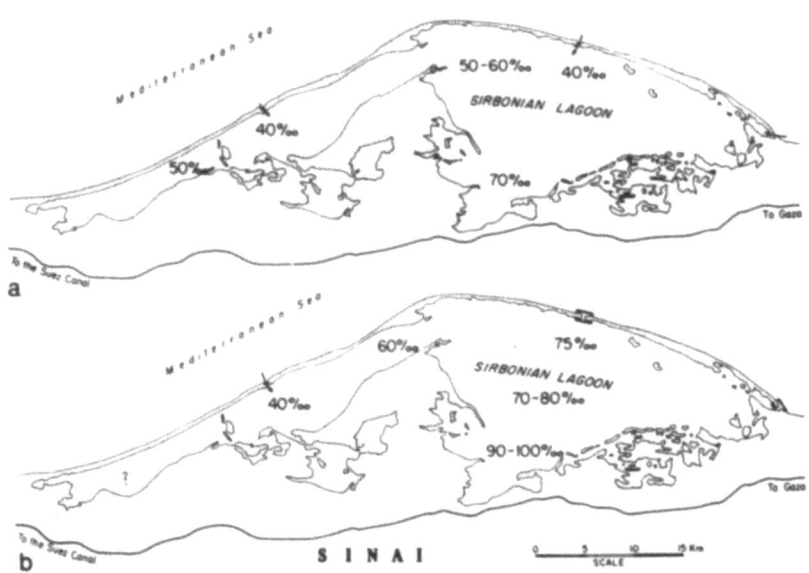

Fig. 5.10. Bardawil lagoon (= Sirbonian lagoon) in northern Sinai. Salinity values are indicated for two situations: a) when the artificial openings are maintained and b) when they are closed with sand (from Por, 1971b).

maximum depth of this shallow lagoon is only 3m. The lagoon communicates with the open sea through an opening in the extreme east and two more, artificial openings. These have to be dredged open from time to time. When the contact with the sea is open, salinities range from 41 to 70 ppt. Bardawil is a typical metahaline lagoon of the Mediterranean. The sedimentary bottom is monotonously covered with cordgrass *Ruppia maritima*. The zooplankton is basically marine, with the copepods *Acartia latisetosa, A.clausi* and *Euterpina acutifrons*, the interesting euryhaline cladoceran *Bosmina coregoni maritima* and the mysidacean *Diamysis bahirensis sirbonica* predominating. In the macrobenthos there are several salinity-resistant mollusks, such as *Pirenella conica, Cerastoderma glaucum, Brachidontes variabilis, Cerithium scabridum, Mactra olorina* and *Angulus valtonis*. The last three are Lessepsian migrants. Among the 41 species of fish reported from the Bardawil, no less than 13 are Lessepsian migrants. Important fishery items in the lagoon are *Sparus aurata* and *Dicentrarchus labrax* as well as several species of Mugilidae. In general, the Bardawil lagoon can be considered as an important "stepping stone" for the Lessepsian migrants, before spreading to the open Mediterranean.

The other four Nile lagoons are well investigated in their chemical aspects (see bibliography with El-Wakeel (1984) and general review by Serruya and Pollingher (1983)). Manzalah, with 1450 square km is the largest; salinities decrease from about 20 ppt near the opening to the sea, to 1.5–2.0 ppt in the more distant parts. Unfortunately, besides the fish fauna nearly nothing is known about the biota there.

Very little is known about the living world of Lake Burullus. It is a low-salinity lagoon and *Tilapia* is the most important fish in the catches.

Lakes Mariut and Edku are better investigated, owing to Steuer (1942), Elster and Vollenweider (1961), Ezzat (1972), Samaan and Aleem (1972) and Saad (1974). Mariut has a maximum salinity of only 10 ppt, while Edku, open to the sea, reaches up to 15 ppt. The vegetation of these lagoons is primarily of *Potamogeton pectinatus* accompanied by *Ceratophyllum demersum*. The zooplankton contains many freshwater species like the cladocerans *Moina dubia* (= *M. micrurum?*), *Daphnia* sp., and *Diaphanosoma excisum* and the copepods *Thermocyclops galebi* and *Mesocyclops* sp. There are, however, also marine plankters like *Acartia latisetosa* and *Diamysis bahirensis*. In the benthos, marine-estuarine forms predominate: the prawn *Palaemon elegans*, species of the amphipods *Gammarus* and *Corophium*, the harpacticoid *Canuella sp.* (*C. perplexa?*). The macrobenthos contains the barnacle *Balanus improvisus*, the polychaets *Phycopomatus enigmaticus* (Fig. 5.11), *Nereis diversicolor* and the hydroid *Cordylophora caspia*. The fish fauna is equally of mixed origins.

Lake Timsah on the Isthmus of Suez, besides having direct historical connections with the Nile, was in contact with Lake Manzalah, through the intermediary of the lagoon of Ballah. Since the times of the Cambridge Expedition in 1924 (see Por, 1978) we have little knowledge about the tribulations to which Lake Timsah was subjected. Surface salinities in the 10 m deep lake may be as low as 20 ppt

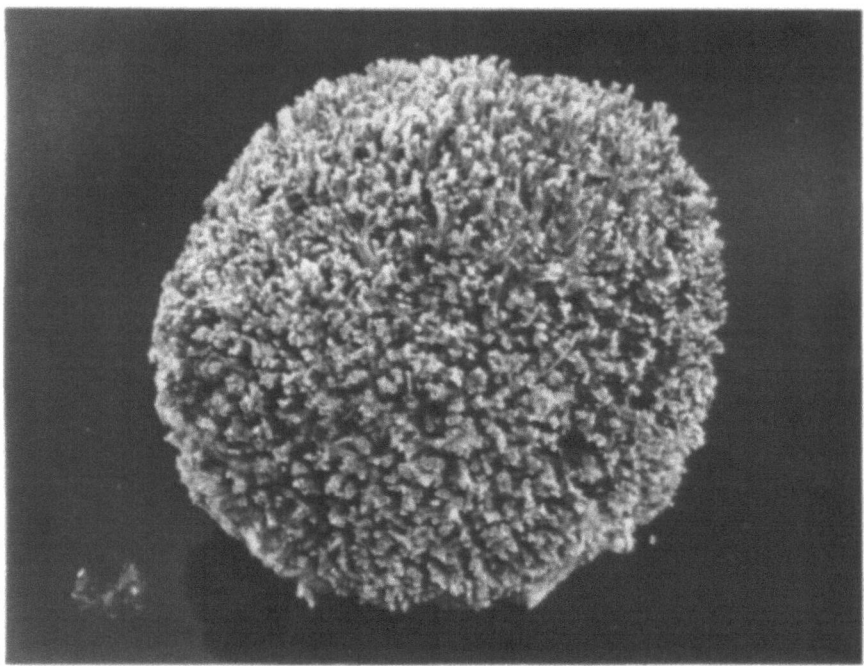

Fig. 5.11. Phycopomatus enigmaticus, Halmyric Serpulid Polychaete (orig.).

when there is freshwater influx and about 43 ppt in the dry summer. The waters exhibit a strong saline stratification for most of the year. Some indications of the older, original Nile-lagoon population are the growths of the freshwater plant *Potamogeton pectinatus* and of the halmyric serpulid polychaete *Phycopomatus enigmaticus*. The rest of the population seems to be by now typical for the Suez Canal system, though with a much larger admixture of Mediterranean species than the neighbouring Bitter Lake.

Without representing anything spectacular or outstanding from the biogeographic point of view, the Nile lagoons and especially Lake Edku and the Bardawil lagoon are the only large halmyric waterbodies along the Levant shores and as such are of a considerable conservational importance.

5.6. Residual brackish estuaries

Perhaps the most disturbed aquatic environments of the Levant are the estuaries of the coastal streams of the Mediterranean shore. The amount of water they carry decreases from north to south as they gradually turn into intermittent flood streams. At present, the southernmost stream still perennially flowing is Nahal Besor in southern Israel. Along the same geographical gradient, the permanently flowing

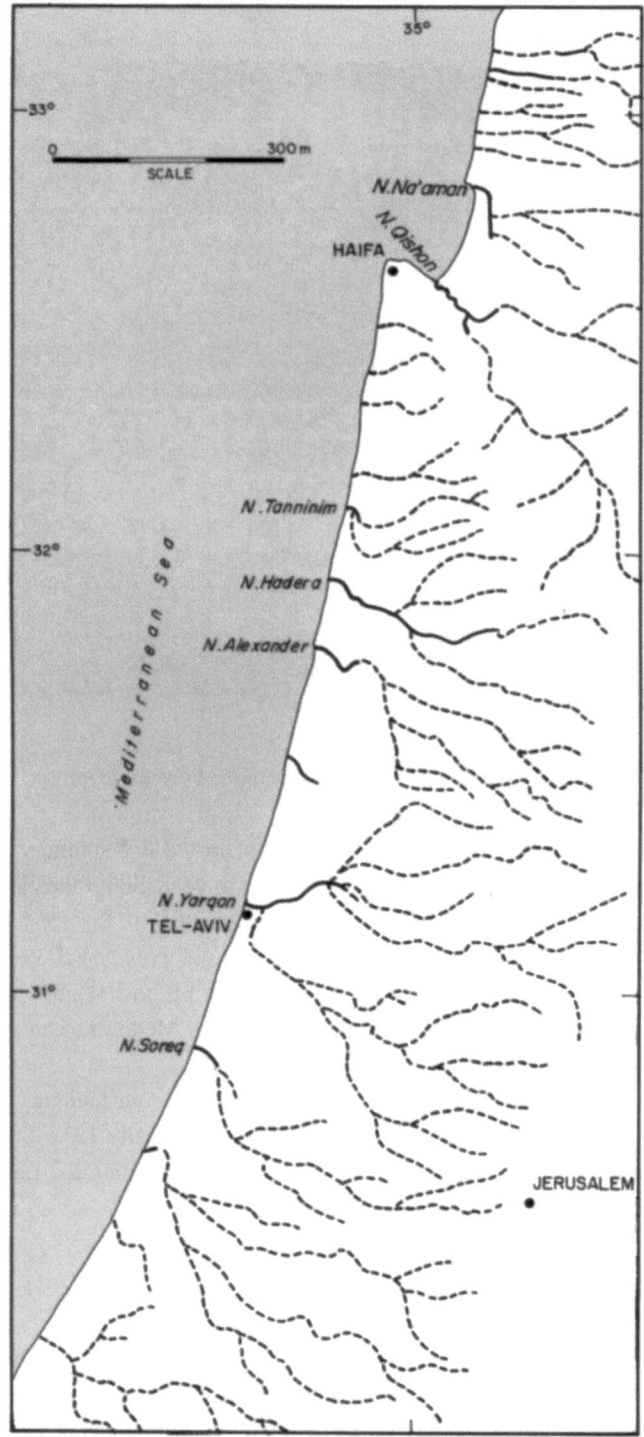

Fig. 5.12. Coastal streams of Israel before recent human impact. Continuous line indicates perennial, dashed line temporary streams (after Atlas Israel, 1957).

stretch becomes shorter and shorter (Fig. 5.12). In recent times, water being taken from the upper courses, the coastal streams are further prejudicated. The instances in which they are used as sewage collectors represent the final indignity inflicted upon them (see below).

Some rather important streams of the Israeli Mediterranean coast have been turned into sewage outlets, like Nahal Naaman, Nahal Qishon, Nahal Alexander and Nahal Hadera. The most important stream of the Israel coast, Nahal Yarqon has been pumped dry completely. Though we do not have data about the lower reaches of the coastal rivers in Lebanon and Syria, their fate is identical, even if somewhat relieved because of the more abundant run-off.

All the coastal streams have small estuaries, although during the dry season they are usually separated from the sea by sand bars; the sand pits are opened during the winter flow or during occasional storms.

Furthermore, the ridge of aeolanitic sandstone which accompanies the coast had the tendency to obstruct the flow of the small streams. When this occurred, large areas behind the ridges or the dunes became oligohaline swamps without drainage to the sea. The largest of these swamps was the Kebara swamp, between Mount Carmel and the coastal kurkar ridges. The swamps made the coastal road, the " Via Maris", impracticable and the Romans had to dig artificial outlets through the kurkar ridges. The swamps contained probably brackish waters and a series of lowland springs which is why the small streams are oligohaline to this day. In the 1930's most of the coastal swamps were drained, but at the same time water started to be pumped from the springs and rivers and pollution started, too.

The estuary of the Qishon became heavily polluted during the 1940's and the Naaman, Alexander and Hadera in the 1960's. Fortunately, some of the information about these estuaries exists, even if unpublished.

Today the only permanent and strong-flowing coastal stream is Nahal Tanninim, the "Crocodile River" which has been spared, first because of its brackish waters and thereafter being declared a Nature Reserve. A few smaller streamlets and estuaries, like N. Oren and N. Daliya still have short stretches of flowing and relatively clean water (Fig. 5.13).

Some work was done in Nahal Alexander prior to its death by pollution (Pizanty, in litt.). The estuary of the small river was characterized by large colonies of the serpulid *Phycopomatus enigmaticus* as well as colonies of the bryozoan *Victoriella pavida*. There were large populations of *Nereis diversicolor* and of marine-estuarine benthic copepods like *Canuella perplexa*. From the sea, the edible blue crab *Callinectes sapidus* as well as eels and grey mullet entered the estuary. *Phycopomatus* was also reported from the Naaman estuary.

Nahal Tanninim has been recently investigated in more detail by Almeida Prado-Por et al. (1981) and especially by Herbst and Mienis (1985). This stream presents chlorinity values of 0.45 to 1.25 ppt. Its fauna is a mixture of oligohaline species and of freshwater species. The halmyric stock found consists first of all of the higher crustaceans *Diamysis bahirensis hebraica*, *Cyathura carinata*,

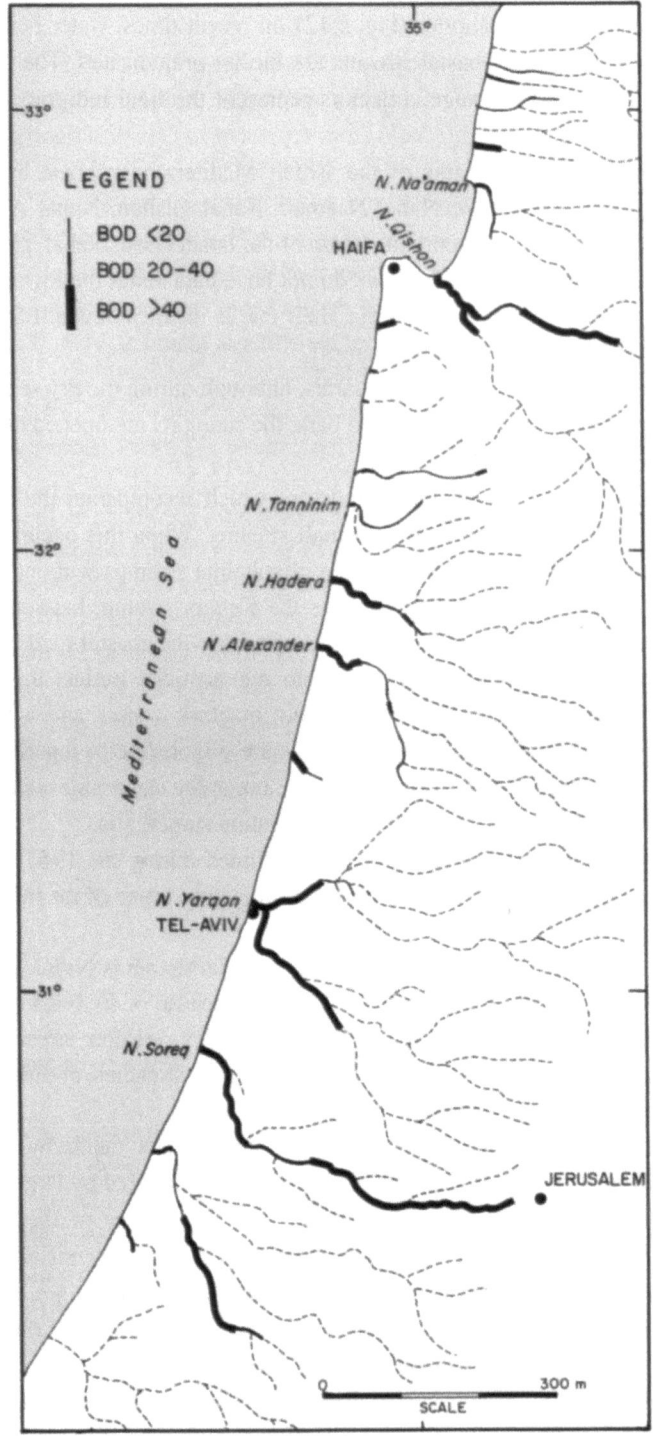

Fig. 5.13. Coastal streams of Israel modified by sewage outflow. Biological Oxygen Demand (BOD) is indicated (after Atlas Israel, 1985).

Sphaeroma hookeri, Corophium orientale, Echinogammarus foxi and *Palaemon elegans*. Among the benthic copepods, the above-mentioned *Canuella* as well as *Onychocamptus mohammed* are common.

As shown by Herbst and Mienis (op. cit.) several of the halmyric crustaceans of the Tanninim also occur in the small estuaries of other coastal streams or did occur before pollution. In general, it seems that the population of the estuaries of the Mediterranean coast was a typical pan-Mediterranean one. An outstanding absentee is the estuarine calanoid *Calanipeda aquae-dulcis*, very frequent in the western Mediterranean estuaries as well as in the Black Sea. *Diamysis bahirensis* is a typical halmyric species complex of the eastern and southwestern Mediterranean. The subspecies described from the Tanninim (Almeida Prado-Por, 1981) seems adapted to oligohaline waters, whereas the subspecies described from the Bardawil lagoon (p. 102) is adapted to metahaline waters.

This is perhaps the place to mention Wadi El Arish with its huge catchment basin, which drains more than half of the Sinai Peninsula to the Mediterranean. Today the totality of the El-Arish drainage is temporary, and permanent waterbodies do not form even at its opening to the sea. There is no doubt whatsoever that during wet phases in the past Wadi El Arish could have carried permanent flow and could have served as an extremely important stepping-stone for the longshore distribution of the brackish and freshwater biota.

Typical freshwater components of the coastal streams of Israel will be mentioned below in connection with the general limnological discussion.

6. The Continental Waters

6.1. Limnology in the Levant

A look at the map of the Levant shows a dense network of streams and rivers. This is a deceiving image. In fact we look at a hydrological skeleton with little flowing life in it. The "mighty Jordan", certainly the most famous river in the world, is something of a limnologic hoax: a poor desert creek, fighting for its existence during the heat of the summer.

The only large river system with a fluvial continuity and stability in the geological dimension, the Euphrates-Tigris basin is strictly speaking outside the Levant. The tectonic phenomena of the rifting, which severed the fluvial continuity between the Levantine watersheds and the Euphrates, created a series of north-south rivers. These, however, confined within the Rift Valley, had a very troublesome history: The Jordan, Litani, Orontes, Kara-Su and Afrin repeatedly or permanently ended in endorheic terminal lakes fragmented into discontinuous sequences or drained to the Mediterranean through short-lived outlets in the coastal mountain ranges.

Weulersse (1940) wrote: "L'Oronte est une creation de la tectonique; ici la vallée a crée le fleuve". Without tectonics, there would be no Orontes and no Jordan.

The tectonics of the "shear" or "transform" movements in the Levantine Rift Valley created according to Garfunkel (1981; 1986) a series of "rhomb-grabens" (called "rhombochasms" by Girdler, 1984) which became permanent sites of the lake basins. According to Garfunkel, such are the grabens of the Dead Sea, of Lake Hula, of the Ghab and of Lake Amiq. (Fig. 2.1). Three such grabens are also situated in the Gulf of Aqaba.

While there was a certain constancy of standing water within the four major grabens, the limnological evolution of the many lakes which dotted the Levant was even more fluctuating than those of the rivers. They expanded or shrunk into extinction with the continuous rhythm of the climatic pulsations of the Pleistocene; they deepened through tectonic subsidence or became obstructed by repeated basaltic outflows; finally, they experienced the impact of the most ancient agricultural civilizations. Of the contemporary lakes of the Levant, one is hypersaline (the Dead Sea), one is on the limit of being brackish (Lake Kinneret), some are man-made (Lake Qarun and Lake Homs), two have been turned recently into farm lands (Lake Hula and Lake Amiq). The rest, not counting one karstic lakelet (Birket

Yamouneh) and one crater lake (Birket Ram), are seasonal lakes or mere pans of salty water.

The majority of the many springs of the Levant bear the mark of the climatic fluctuations of the area. At humid times they produce permanent streams and rivulets and when aridity sets in they retreat uphill and even temporarily disappear underground. Man has been tormenting them since time immemorial by tunneling into them, sequentially puncturing their aquifers (with the "qanat" system) or at best capturing every single drop into impressive aquaducts or into irrigation networks. In the oases, the life-giving springs have for millenia been trampled on by caravans or sealed off by conquerors. The aquifers themselves bear the marks of the complex geological history of the Levant: they are a mixture, sometimes stratified, sometimes side-by-side, of fresh, brackish, saline and hot juvenile waters in all the possible proportions.

The main base of limnological life in the Levant consists of several huge karstic springs, real resurgent rivers which erupt above-ground with their full discharge, often under considerable artesian pressure. They are not seasonal and the multiannual changes in the pluviosity are only slowly influencing their output. The rivers they form are to a considerable extent freed of the dependency on the unpredictable surface run-off.

First among them are the springs of the Jurassic aquifer of the high mountains of the Antilebanon, such as the Dan (headwater of the Jordan, Por et al., 1986), the Zarqa, headwater of the Orontes (Nahr Assi) and En Fidje, the source of the Barada, the feeder of the oasis of Damaskus. These sources, tapping a very deep freshwater aquifer, do not exhibit any seasonal or short-term fluctuation in their discharge. They are the safeguards of the ultimate survival of the freshwater biota in the Levant.

The aquifer of Mt. Hermon supplies no less than 750 million cubic meters per year, of which the spring of the Dan, alone takes one third. Consequently, Dan alone supplies about half of the minimal summer flow of the Jordan (measured at its exit from Lake Kinneret).

Weulersse (1940) speaks of such springs as "un fleuve qui nait de lui même". They are all situated within the talweg of the Rift Valley rivers or in its immediate vicinity. For the Orontes alone, Weulersse counts more than 20 such artesian springs, from Zarqa, the real headwater (12 cubic meters/second) to Daphne, in the far north, near Antakiya. In the short stretch of the Ghab valley there are several artesian springs, with a total output of 15 cubic meters/second.

Less impressive, but still remarkably abundant are the springs which feed off the rainfall on the huge basaltic areas like Gebel Hauran and appear at the edges, where basalt meets impermeable senonic limestone. Such are, for example, the spring of Mzarib, the headwater of the Yarmukh, or the springs of Qa el Azraq in the Jordanian desert.

With the exception of these oases of stenothermic, well-oxygenated and totally fresh waters, all the other waterbodies of the Levant show strongly fluctuating limnological parameters. They frequently turn salty, reduce their flow and as a consequence warm up and become oxygen-depleted. The only feature which is invariably characteristic to all the waters in the region is the high carbonate content and the considerable hardness of them.

As a consequence the inland waters of the Levant are dominated by biota which are resistant to environmental changes, especially to salinity fluctuations, high temperatures and low oxygen content. Biota which have the capacity of surviving difficult periods through resting stages, by going underground or by means of aerial dispersal are at an advantage. The so-called secondary freshwater fauna predominates over the primary one. For the aquatic biogeographer, the Levant is a difficult area.

With the exception of the subterranean species and a few other cases (fishes!), the Levant waters have not had the time to produce endemic species. Areas of endemic species tend to include the Mesopotamian province, the large basin of the Euphrates. Many endemics are, or will probably turn out to be young-aged subspecies. Yet, one has to make use of every bit of information in order to sort out the primary biogeographic connections from the adventive, secondary ones. Historical biogeography in the Levant often explains much less than ecological biogeography: fluviatile contacts and limnological stability play a lesser role than environmental adaptability and a good capacity for overland dispersal.

In the Levant, where mastery of the water sources was often a prime reason for wars, the limnological muses are still silent. The inland water fauna of the province is largely unknown since the political barriers are often closed. Therefore many species, newly described, are proclaimed as "endemics" of a certain basin, since their presence on the other side of the political border is unknown.

Because the need of water is great and scientists are always late, the restricted areas of many species are man-made artifacts. Taxonomic information is on an average scanty. There is good information on fishes and mollusks; decapods are reasonably well known, but amphipods and isopods much less. Copepods, cladocerans and especially ostracods are poorly known for the Levant as a whole. Among the aquatic insects, orders like Odonata, Ephemeroptera, Plecoptera, Trichoptera and Hemiptera are well known; others, like the orders Diptera and Coleoptera are not. Knowledge on watermites is practically non-existent. Similar is the case of the nematodes and rotifers, not to speak of gastrotrichs and tardigrades. Planarians and oligochaetes are locally better known.

Because of all this, the treatment of the inland water biogeography of the Levant will be spotty and preliminary. The selection of the main issues through which I want to give a biogeographic picture as complete as possible can be open to criticism, too. We shall try to discuss the complicated subject by emphasizing the biogeographic origin of the Levantine inland water biota. Since I consider the euryhaline biota as being the geologically oldest, they will be discussed first. Then I shall proceed to discuss the oldest autochthonous freshwater fauna, the Mesopotamian one; this will be followed by the Ethiopian connection and finally by a discussion of the recent Palaearctic influence. Where information is available, several waterbodies will be singled out for discussion in the place which seems appropriate to us. Finally, the special case of the fauna of the ephemereal and man-made water bodies will not be neglected.

6.2. The salty waters of the Jordan Valley

Instead of discussing the Jordan as a whole, for reasons exposed below, we shall now consider the lower reaches of this river, from Lake Kinneret to the south.

The fresh waters of the Jordan and of several tributaries are literally cascading down into the deepest continental tectonic chasm of the globe: Lake Kinneret (also called Lake Tiberias or Sea of Galilee) has a lake level around –210 m. After leaving the lake, the Lower Jordan meanders even deeper to the –400 m level of the Dead Sea.

In this peculiar environment, the waters of the Jordan encounter the fossil salt layers of lake Lisan (see below) and a wealth of mineral springs, products of the active tectonism of the Rift Valley. As a consequence, the waters tend to become salty and the ionic composition of the dissolved minerals deviates much from the marine ratios. This trend of salination increases from north to south. In the time scale of the Pleistocenic history, it changed at short millenial rate. The salinity climax was reached during 70,000–18,000 B.P., when the hypersaline Lake Lisan covered the whole Lower and Middle Jordan valley (see p. 35).

A glimpse at a low salinity situation is given by the fauna of Lake Ubediya, which existed around 600,000 B.P. (Horowitz, 1978) or 1.4 million B.P. (Tchernov, 1987). According to Horowitz (pers. comm.), the Jordan Rift Valley contained during the Pleistocene three separately evolving lacustrine basins: the southern one, corresponding in part to the present Dead Sea, the central one, less permanent, of which the Kinneret is a young offspring and the northern one, surprisingly constant, which contained Lake Hula.

According to Horowitz, there are reasons to believe that the Hula never became part of a unified lake. Horowitz also believes that the large-extension Lake Lisan had possibly several earlier and less known antecessors.

The origin of the salinity in the Lower Jordan valley is diverse. Besides the high evaporational water loss, leaching of the salty Lisan marls is also an important source. But probably most important are the many highly mineral and sometimes hot springs on the shores of the Kinneret, the Lower Jordan and the Dead Sea.

In a series of publications Mazor and his co-workers studied the origin, age and composition of these springs (see summing up by Mazor, 1978). It is suggested that the mineral springs along the Rift Valley have the same origin and composition: they are waters of the Pliocenic or older marine gulf which, after being trapped, sometimes at considerable depth, are driven out by tectonic stresses. On their way they dissolve minerals from the local rocks and before opening to the surface, they mix with various proportions of meteoric water. These waters, called by Mero (1978) "N-T waters" ("Noit-Tiberias waters") contain non-marine, aberrant mineral ratios, like the as yet unexplained, low Cl/Br ratio. It is assumed that the total mineral content of the undiluted N-T waters is slightly higher than that of the original sea water, probably due to evaporation (Gat et al., 1969). The less diluted N-T springs have a salinity of around 30 ppt (Tiberias hot springs). The saline springs along the shore of Lake Kinneret supply about half of the annual chloride input of the lake. As a consequence, the salinity of the lake was around 0.5 ppt before the recent desalination efforts.

Recently, Ehrlich (1986) has discovered a brackish-saline diatom assemblage in cores taken from the bottom of Lake Kinneret. These diatom species are similar to those described by Begin et al. (1974) from the at least 18,000 year-old deposits of the ancient Lake Lisan. In Ehrlich's interpretation, as late as 5000 years ago, Lake Kinneret was much smaller and shallower and surrounded by brackish spring-fed swamps. Since that time, the lake bottom subsided by at least 10–12 m to reach its present maximal depth of 42 m.

Several salt springs of the Bet Shean valley and of the Lower Jordan are also of the N-T type. Along the steep western escarpments of the Dead Sea, three basic types of springs are encountered: 1. The springs which emerge several tens of meters above the Dead Sea level are freshwater springs with temperatures corresponding to their location (e.g. 27°C in the spring of En Gedi); 2. the springs which appear several meters above the lake level are saline springs of the N-T type; they have slightly brackish salinities, like most of the En Fashkha springs; 3. finally, countless small springs emerge near the shoreline of the lake and are very salty and sometimes sulphurous. According to Mazor (1978), these springs represent various mixtures of N-T, fresh and Dead Sea water. Some of these springs, like Hamei Zohar, are hot.

We do not know much about the water chemistry of the springs on the eastern side of the Dead Sea. The case of the spring complex of Zerqa Main is, however, most peculiar: according to Bender (1968) it is composed of about 18 springs with salinities of around 20 ppt. Temperatures measured in these springs reach 63°C. The highest of these springs is situated 470 m above the Dead Sea level and descends as a hot waterfall.

The present Dead Sea is a concentration basin of all the mineral springs of the Rift Valley of the Jordan and indeed a huge evaporative system. The process which led to the present salinity of around 320 ppt is the result of a very long period of saline accumulation. The expansive Dead Sea, which under the name of Lisan Lake occupied the whole valley up to the southern part of the present Lake Kinneret during the latest Pleistocene, probably had salinities around 150 ppt. The possible reasons for this expansion have been discussed above (p. 35). Nonetheless, as summarized recently by Begin (1986), Lake Lisan had a rich euryhaline flora of diatoms in its northern half. There are even deposits of diatomite from that area. Rosenfeld (in lett.) found several species of ostracods in the lake sediment, such as

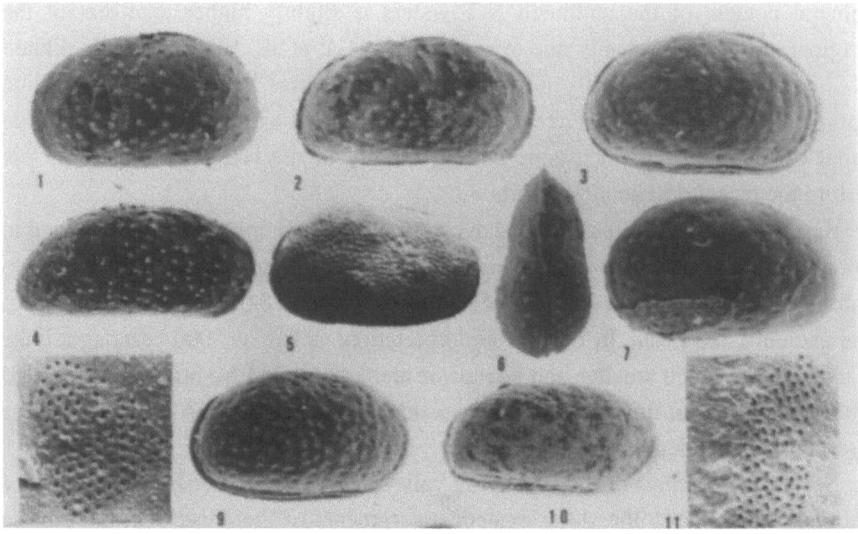

Fig. 6.1. Cyprideis torosa halmyric Ostracod. Fossil shells from the Pliocene Bira formation in the Jordan Valley. Different pore structures are indicated, corresponding to different salinity values (from Rosenfeld et al., 1981).

Cyprideis torosa, Candona sp. and *Ilyocypris sp.* Fossil remnants of *Aphanius* were found, too.

It is interesting to mention that Rosenfeld et al. (1981) found a similar faunula of ostracods in the deposits of the Upper Pliocene Bira lagoons (see p. 14). Salinities inferred by the pore structure of those Pliocene *Cyprideis* shells are between 10 ppt and 100 ppt (Fig. 6.1). Snails of genus *Hydrobia* were also frequent in the Plio-Pleistocene waterbodies of the Jordan valley. Thus the whole euryhaline fauna found there today has probably a Pliocene age, if not older than that.

We shall briefly discuss the biota of the Dead Sea, of the rift valley springs, of the Jordan and of Lake Kinneret, with emphasis primarily on the euryhaline biota. The other component present in these waters, namely the old or more recent Afrotopical or Ethiopian one, will be discussed separately, in a broader context.

6.3. The Dead Sea or nearly so

One cannot avoid dedicating a few words to this most saline water body. The sheer dimension of this lake makes it necessary: Early in this century the Dead Sea was 78 km long and had a maximum width of over 17 km. The maximum depth, as quoted in the world literature, was given as 395 m. Recent measurements failed to find more than 330 m. Traditionally, the Dead Sea was considered to be abiotic, or

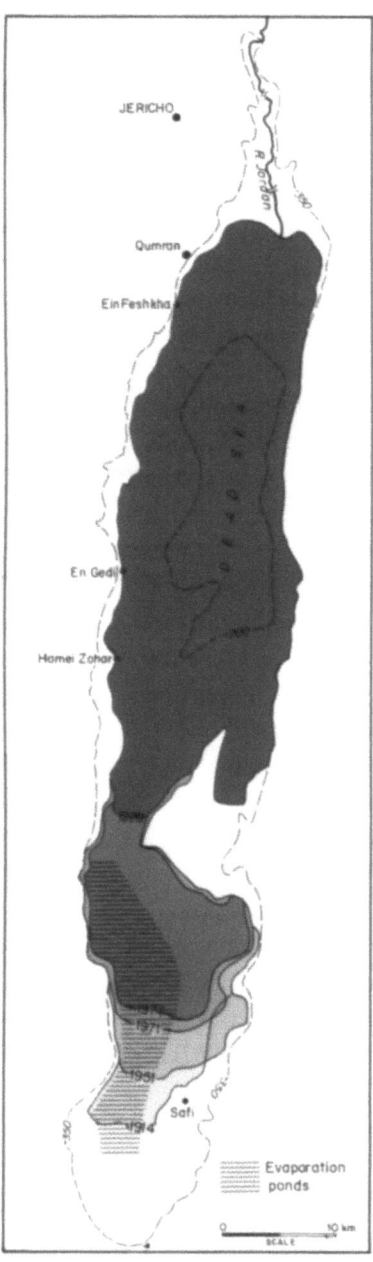

Fig. 6.2. Receding shorelines of the Dead Sea. Dates are indicated. Hypsometric curve of −350 m indicates the assumed maximum historically recorded lake level (after Klein, 1982).

at least azoic. Volcani (1936, 1940) was the first to demonstrate that bacteria and unicellular algae live in this lake. At about the same time the Rumanian protozoologist I. Lepsi claimed to have found rhizopods and ciliates there.

After the more recent work by Kaplan and Friedman (1970) it is evident that the only alga is *Dunaliella parva* and that there are several bacteria; among them the most frequent and haloresistant are two types of *Halobacterium*. Several bluegreen algae and protozoans have been hatched from cysts found in the sediments of the lake.

The relatively high content of Mg and possibly also of some other ions is held responsible for the lack of any metazoan life in the waters of the Dead Sea.

As documented by Klein (1982), the Dead Sea has continually regressed since 1929 when its level stood at −391.5 m. By 1979 the level was 12 m less, i.e. −403.5 m, and the southern shallow basin dried out. In all the southern shore of the lake receded by some 30 km (Fig. 6.2). Klein (op. cit.) considered that the reason for this regression was purely climatic and that in the past, when pluviosity was low, similar low levels were reached. It seems, for instance, that in the classical times, if one takes the description by Josephus Flavius (first century A.D.) and during a later phase represented by the Madaba map (sixth century A.D.) the lake level was similar to the present regressive one (Neev and Emery, 1967). There is also a description of a similar situation from 1818. Other authors believe that the recession of the lake levels is also due, at least in part, to the intensive use of the Jordan waters for agricultural purposes and the consequent lowering of the water input into the Dead Sea. The present rate of shrinking of the Dead Sea is expressed by an average annual lake level lowering of 30 cm (Fig. 6.3).

As a consequence of the decreasing water levels, the Dead Sea has undergone a remarkable limnological revolution: In 1967 Neev and Emery divided the lake into a 40 m deep epilimnion with a salinity of 290 ppt; a pycnocline zone between 40 and 50 m depth and then a hypolimnion with 325 ppt salinity down to the 330 m deep bottom (Fig. 6.4). Due to the retreating water levels, the epilimnion has been literally peeled off and finally in the winter of 1977 (Beyth, 1980) general mixing of the water column occurred at a homogenuous salinity of 340 ppt. Even some amounts of oxygen reached the deep bottoms. The hypolimnic waters once called "fossil waters" by Neev and Emery (1967) reached the surface probably for the first time in centuries.

Indeed this occurrence did not have any positive effect on the restricted biotic diversity of the Dead Sea: it remained the only known large waterbody in which no animal grazing is linked into the process of organic recycling.

Klein (1982), who studied the historical fluctuations of the Dead Sea levels in the broadest aspect, considered the present level decrease as normal. Indeed, during the last century B.C. the Dead Sea level rose to a level of − 330 and submerged the buildings of the Essenian community of Qumran. Klein succeeds in dating this event around the year 33 B.C. A rapid fall in the lake level followed and within less than one century it reached −395 m. According to Klein (op. cit.) the high levels at

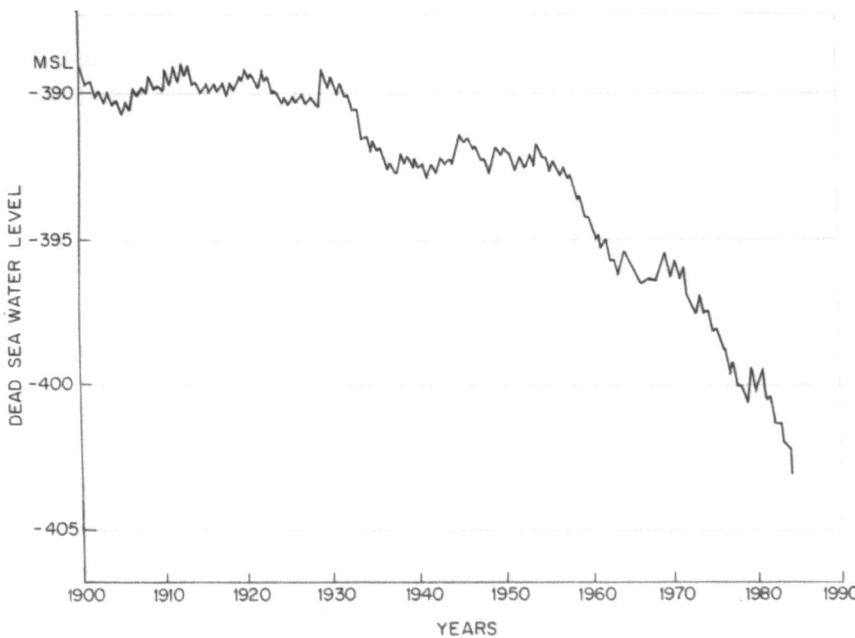

Fig. 6.3. Fall of the Dead Sea level in the present century – MSL = Mediterranean Sea Level (after Klein, 1961 and Hydrological Service of Israel).

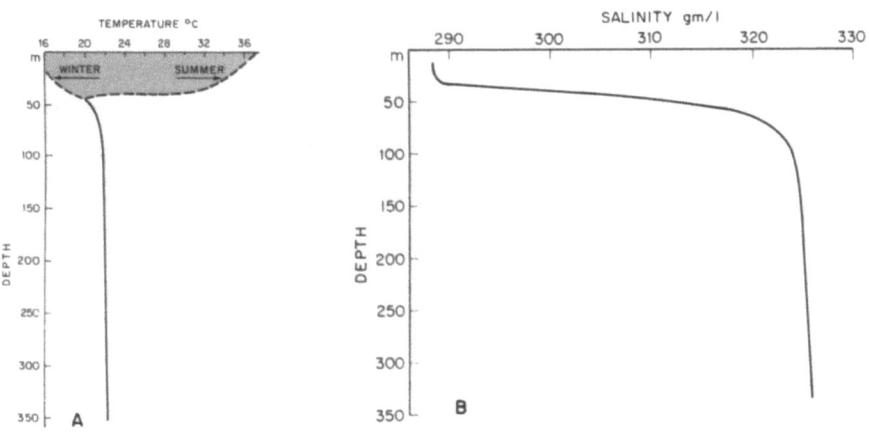

Fig. 6.4. Dead Sea hydrology before the 1980's. (A) Temperature; (B) Salinity (after Neev and Emery, 1967).

the beginning of this century were reached through a rapid lake level rise in the nineteenth century.

The project of leading Mediterranean waters into the Dead Sea depression for the sake of energy production has led to several studies, before being shelved in

1986. A series of simulation experiments performed by Dimentman showed the diversity and limits of survival of Mediterranean and local organisms in various mixtures of Mediterranean and Dead Sea water. From these experiments it became evident that even an admixture of 10% of Dead Sea water to the Mediterranean water became lethal for all marine organisms. Only a few Protozoa, brine-flies (Ephydridae) and the ephemeropteran *Cloeon dipterum* resisted this mixture.

Today, the Dead Sea is a deeply disturbed environment. Its southern basin is artificially filled through a channel coming from the deep northern basin. The channel supplies water for a dozen hotels and spas and especially for the Israeli and Jordanian potash- and bromide extraction plants. Along the shores of the northern basin, hundreds of meters of shore are uncovered and turned into salty badlands. However, some springs previously covered by the Dead Sea are accessible now: they are yielding a rich subterranean crustacean fauna which is being investigated at present by Dimentman and Bromley.

6.4. Saline oasis springs

As mentioned earlier, the perimeter of Lake Kinneret, of the Lower Jordan and of the Dead Sea contains an extremely large number of springs of different salinities. This should be understood in the widest sense of the term: in fact, the area of the salty springs corresponds more or less to that of the old Lake Lisan. Consequently the area includes also the salty springs of the Bet Shean valley-triangle, some 10 km west of the Jordan and especially the springs south and southwest of the Dead Sea as distant as 50–70 km (En Yahav and En Avedat). On the escarpments of the Rift Valley, there are many freshwater springs which emerge above the level of the old Lake Lisan: they are freshwater springs and their fauna harks back to different biogeographic origins.

The different spring-groups on the shores of the two lakes contain as a rule a mosaic of springs with different salinities, without even any difference in altitude between them. For instance in the En Fashkha complex on the shore of the northern Dead Sea, salinities range from 2.5 to over 50 ppt (Mazor and Molcho, 1972); in the nearby Enot Samar complex, there are even a few completely fresh springs. In the En Fuliya complex on the shore of Lake Kinneret, the reported variation is from 284 to 1,400 mg Cl/l; in the Tabgha group also on Lake Kinneret, it is from 1,318 to 2,325 mgCl/l (Mazor, 1978).

The biota found in these springs are either relic marine species, or brackish, halmyric and otherwise euryhaline freshwater species. Our assumption is that since the Miocene marine transgression, the Pliocenic marine lagoon situation and throughout the many arid phases of the Pleistocene there were always halmyric waters in the Lower Jordan and Dead Sea valleys.

The biota behaved accordingly. The diversity of saline spring refugia in the area as well as of the different groundwater mixtures supplied the basis for alternative advance and retreat as salinities in the large bodies of open water changed (Por, 1968b). During the last 18,000 years there existed a fluviatile continuum along the Middle and Lower Jordan, which allowed for smooth repopulation of tributary springs and brooks. The Dead Sea on the other hand, was an almost impenetrable hypersaline barrier for the last 18,000 years. The 70,000 preceding years, while the saline Lake Lisan was in existence, were hardly more proptious.

Therefore an interesting bioegeographic discontinuity exists south of the spring oasis of En Fashkha: faunal influx through the Jordan was and is still impossible. For instance, the fishes of genus *Tilapia* advancing south from the River Jordan are confined to the Fashkha oasis. South of these springs only relic populations and readily spreading insects and other invertebrates can be found. The few freshwater streams of the western side of the Dead Sea are short and have small catchment areas: during the high water levels of the Lisan Lake (about 200 m higher than that of the present Dead Sea), many of the streams were drowned. Consequently, none of these streams contains any freshwater fish fauna. The river-sized streams of the eastern side of the Dead Sea, like for instance Wadi Mujib or Es Safi, with large catchment areas, have fish faunas of their own.

The oldest and probably most stenohaline element in the Rift valley springs are the subterranean marine relic crustaceans (see above, p. 6). These species are associated with the N-T type waters (p. 113), mixtures of the Pliocenic lagoon water and meteoric water (Mazor, 1978). This fauna is characterized by the well-known four species: *Typhlocaris galilea, Monodella relicta, Typhlocirolana reichi* and *Bogidiella hebraea*. Intensive research done in the Dead Sea and Negev areas by Herbst (1982), Herbst and Dimentman (1983) and Herbst and Bromley (1984) failed to find further subterranean relic crustaceans. Recently, however, the receding level of the Dead Sea exposed springs which yield a rich subterranean fauna (Dimentman, pers. comm.).

The less well-investigated springs of the Lake Kinneret shores might yield surprises, too. Troglobiosis was no doubt an important way to survive the frequent hydrological changes in the Rift Valley. For instance, a probably endemic harpacticoid copepod *Nitocra balnearia* (Por, 1964b) found in the springs of Hamei Zohar (also the "locus typicus" of *Monodella relicta*) had to hide underground when the Dead Sea waters covered the springs a few decades ago.

In the surface flows of the springs, euryhaline species of halmyric origin are frequent. The case of two amphipod crustaceans is interesting (Herbst and Dimentman, 1983): The genus *Echinogammarus* is represented in Israel by *E. foxi*, which is frequent in the estuarine coastal streams, *E. veneris*, which is found in and around Lake Kinneret and along the Lower Jordan and a third species (not yet described) which is found along the Dead Sea and in the springs of the Negev. Somewhat different is the case of *Orchestia*: the species *O. cavimana* is found in the Mediterranean coastal streams and *O. platensis* along Lake Kinneret and the Lower Jordan (though also in the Judean mountains and the now drained Lake Hula).

The extremely euryhaline fish genus *Aphanius* has an even more complex distributional picture. The more oligohaline *A. mento* is found in the Rift Valley from En Fashkha to the north; the extremely euryhaline *A. dispar richardsoni* is frequent in the Dead Sea springs, south of En Fashkha, while a recently described species, *A.sirhani* lives in the springs of the El Azraq oasis in the Transjordanian desert (Lotan, 1982; Krupp, 1983; Vilwock, 1987). In the Red Sea there is an *A. dispar dispar*, which lives also in the Levant basin; a further species, *A. fasciatus*, lives in the eastern Mediterranean.

These are pieces of information pointing to the fact that an evolution of endemic taxa eventually occurred among the saline water fauna of the Jordan Rift Valley. Even one tricladid planarian *Dugesia biblica biblica* represents a subspecies, eventually typical for the Rift Valley. Another species, *D. salina*, lives in Lake Kinneret and the springs of the Bet Shean valley (Bromley, 1974). Tchernov (1971a) described a hydrobiid snail *Pseudamnicola solitaria* from the springs of the Dead Sea area. Another hydrobiid, *Semisalsa longiscata*, lives along the Israel and Lebanese sea shore (Schütt, 1983) and lived also in the Lower Jordan valley (Tchernov, 1973). The taxonomy of the euryhaline Hydrobiidae is still too incomplete to allow biogeographic conclusions.

The snail *Melanoides tuberculata* is widespread in all the springs of the Jordan rift valley; this is a well-known euryhaline freshwater species of the Old World tropics.

Euryhaline species are also common among the lower crustaceans, such as the ostracods *Cyprideis torosa* and *Herpetocypris salinarum* and the cyclopoid copepod genus *Halicyclops*.

Among the aquatic insects, which are dispersing easily, hydrophilid, dytiscid and hydraenid Coleoptera are common in the salty springs (Herbst and Bromley, 1984; Jäch and Margalit, 1987), equally represented are the Hemiptera (Linnavuori, 1960, 1964) and certain families of the Diptera.

Botosanenanu and Gasith (1971) studied the Trichoptera, an aquatic insect order known to have several euryhaline taxa, too. The genus *Hydroptila* and especially *H. simulans* known from a variety of aquatic environments in the Rift Valley is a good example of this.

Herbst and Bromley (op. cit.) analysed the aquatic fauna of the Dead Sea shores and of the Negev desert and concluded that the highest species richness is found in the springs and at an average chlorinity value of 1.0 ppt. The oligohaline springs of the area are the most stable aquatic environments around. Typical freshwater habitats which result from rain water accumulation, like pools and cisterns, are too unstable or short-lived.

The oasis of El Azraq contains the only natural standing waterbody of the Kingdom of Jordan, a spring-fed lake in the eastern desert of that country. The above-mentioned presence there of a species of *Aphanius* as well as the presence of the freshwater prawn *Atyaephyra desmarestii orientalis* (Kinzelbach, 1987) and of several other aquatic biota, justify the mentioning of this environment together with the salty spring-oases. However, too litle is known about this interesting oasis. There is good reason to believe that El Azraq is an extremely impoverished relic of an ancient Pliocenic connnection between the nascent Jordan rift valley and the Euphrates drainage (Kinzelbach, 1987).

6.5. The Lower Jordan

The Jordan, south of Lake Kinneret is a very young river: it started to cut its vallley again after 18,000 B.P., following the withdrawal of Lake Lisan. For sometime after this a brackish lake still existed in the area of the Bet Shean valley of today.

From the beginning, the Lower Jordan was settled by a euryhaline freshwater fauna. It has very little in common with the more freshwater fauna of the Upper Jordan which empties into Lake Kinneret in the north.

Unfortunately, nothing is known about the original fauna of the Lower Jordan. The changes to which it has been submitted have been numerous and traumatic (Ortal and Por, 1978; Por and Ortal, 1986). With the building of a dam at the outlet of Lake Kinneret in 1933 and the subsequent building of the National Water Carrier using the water of this lake, the outflow of the Lower Jordan decreased by more than 90%. Further damage was done when a saltwater diversion channel brought the waters of the saline springs of the Kinneret directly into the Lower Jordan. In close sequence, water from the major tributary of the Lower Jordan, River Yarmoukh, was gradually diverted, beginning in 1958, into the major irrigational system of the Kingdom of Jordan, the eastern Ghor Channel (Fig. 6.19).

The only economic consideration which granted the suspended animation of the Holy River, was the need of the recent agricultural farms along the Jordan itself. The impact of all of these changes which diminished the annual discharge of the Lower Jordan from 1,100 million cubic meters/year to 100 million cubic meters on the Dead Sea was mentioned above (p. 116). Furthermore, the quality of the Jordan waters changed, becoming much more salty, and polluted by organic sewage and industrial effluents.

Ortal and Por (1978) present data about the biota of the upper stretch of the Lower Jordan, most influenced by the hydraulic changes and by saline, domestic and chemical pollution. Even at present a fair diversity of the Lake Kinneret mollusks and cyclopoids is found. Among these, the African *Afrocyclops gibsoni* was reported for the first time in the area. The harpacticoid copepods are represented by a complex of brackish species, namely *Nitocra incerta, Onychocamptus mohammed* and *Cletocamptus deitersi*. The above-mentioned amphipod *Echinogammarus veneris* is widespread. Other more or less euryhaline crustaceans of freshwater origin, the isopod *Proasellus coxalis* and the prawn *Atyaephyra desmarestii*, were also found in the Lower Jordan.

Recently, the Mediterranean euryhaline prawn *Palaemon elegans* appeared in the Lower Jordan. Today it is widespread and found even near the Jordan outflow

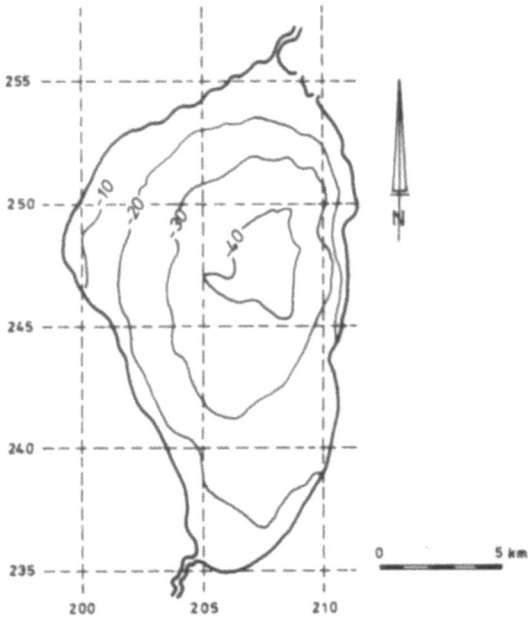

Fig. 6.5. Simplified bathymetric map of Lake Kinneret.

to the Dead Sea (Dimentman, unpublished). One of the possibilities to be considered is that *Palaemon* was accidentally introduced together with the Mediterranean mullet fry which is transferred from time to time from the coastal rivers to Lake Kinneret and adjacent ponds, for the sake of the fisheries in the area.

6.6. Lake Kinneret – the Sea of Galilee

Around 18,000 B.P. Lake Lisan receded. The northernmost corner of it subsisted or even expanded to form the much more diluted Lake Kinneret. As shown above there is also an indication that tectonic deepening of the basin occurred since that time (Fig. 6.5).

The different names given to this only large natural lake of the Levant are mentioned above. Kinneret is indeed a large lake, even by general standards. Information about this lake has been summed up by Serruya (1978). Its length is 22 km and maximum width 12 km; depth, depending on water level is 43 m with a mean depth of 25.6 m. The shores are steep and angles of 5 percent are not uncommon (Serruya, op. cit.). The lake level is deep below the Mediterranean Sea level: it fluctuated in accordance with the annual pluviosity in the historically documented period, from –211 to –209 m. Since 1932, when a dam was built at its southern outlet for hydro-electric purposes, the lake level has been artificially controlled.

Lake Kinneret is an oligohaline lake: the total value of dissolved solids is around 0.6 ppt and occasionally reached over 0.7 ppt. This is slightly above the accepted lower limit of 0.5 ppt of brackish-oligohaline waters.

The Upper Jordan supplies Lake Kinneret with an average of 558 million cubic meters per year: these are fresh waters, with normal ionic composition. The high salinity and specific ionic composition of the dissolved salts in the lake is due to the above-mentioned salt springs which supply over 100 million cubic meters/year (Mero, 1978).

Lake Kinneret is a warm-monomictic lake according to Hutchinson's classical definition: during most of the year (early April to late December) the water column is stratified, the epilimnion with temperatures of up to 30°C and the hypolimnion with temperatures around 14°C; during a short season (January–March) the epilimnion cools and the lake water is destratified and homogeneous at around 13–16°C. During stratification the hypolimnion is anoxic. Some years are on record in which the destratification occurred as late as February and lasted for only a few weeks.

As a rule, the thermocline and the oxycline are situated at a depth of around 20 m, showing a certain seasonal or sometimes a wind- or seiche-induced fluctuation (Fig. 6.6).

It is interesting to mention that warm-monomictic lakes are rare around the globe: they are characteristic of the subtropical climatic belt, where large and deep lakes are a rare occurrence. Were it not for its location in a tectonically active rift valley, Lake Kinneret would have rapidly turned into a shallow, salty swamp. The lake is also on the limit of becoming an oligomictic or even an amictic lake of the tropical type: in such lakes (like the great African lakes) mixing occurs only irregulary or not at all. A few years of higher winter temperature could easily turn Lake Kinneret into an oligomictic lake; such occurrences cannot be ruled out in the subrecent history of this lake.

Lake Kinneret is characterized by a somewhat restricted biotic diversity. This is due to three reasons which, not in the order of their importance, are: 1) The lack of oxygen in the deeper water; 2) the oligohaline salinity of the water; 3) the almost complete absence of submerged and floating vegetation. The lack of vegetation may be due to the steep lake shores, the open and non-sheltered coastline and the strong winds which disturb the lake surface.

Around a depth of 15–30 m, there is a steep reduction in the benthic biomass, i.e. from over 7 g/m to nearly zero. This zone, subject to the fluctuations of the thermocline, has been called by Por (1968a) the "epiprofundal" of the lake. In the deepest "profundal" zone, devoid of oxygen for at least 8 months, there is again an increase in benthic biomass to values of around 1 g/m (Por and Eitan, 1970).

Owing to the average depth of the lake, some 60% of the total benthic biomass of the lake is found in the profundal. However, it is composed only of the anaerobic nematode *Eudorylaimus andrassyi* and the tubificid oligochaetes *Potamothrix heuscheri* and *P. bavaricus* (Por, 1968a; Brinkhurst and Jamieson, 1971). As shown

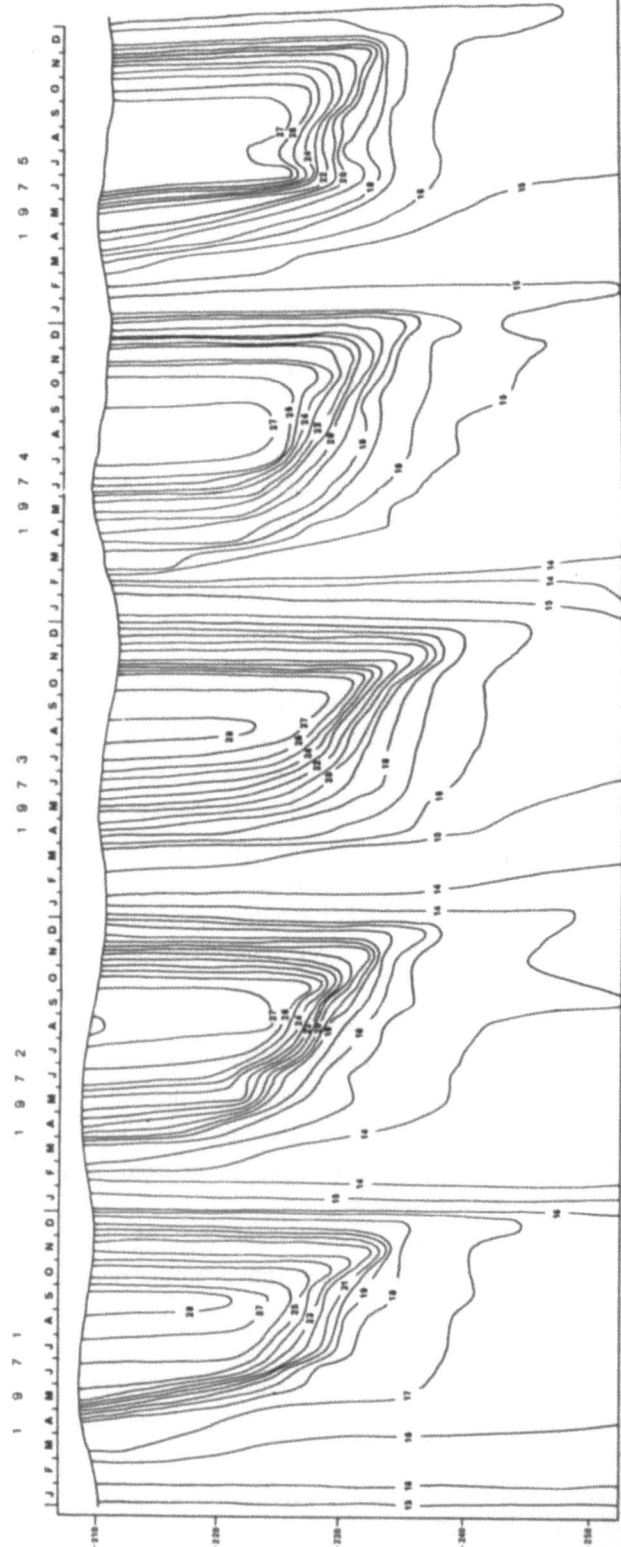

Fig. 6.6. Lake Kinneret thermal profiles for 1971-1975. Lake level fluctuations are indicated (from C. Serruya, 1978).

by Por and Masry (1968) these organisms are extremely tolerant to anaerobic conditions. The nematode may reach populations of many tens of thousands of specimens per square meter. On these grounds Por and Eitan (1970) called on these grounds Lake Kinneret a "nematode lake".

Other organisms appear in the deep water or on deep bottoms only during the short season of overturn. Among them, the most resistant are the chironomids *Procladius choreus*, *Tanypus punctipennis* and *Cryptocladopelma virescens*; they survive in the oxygen-depleted profundal till the month of May (Kugler, 1978).

It is interesting to mention that during the long months of stratification, there is also a slight increase in the salinity of the hypolimnion, at least partly due to the flow of several known and many more unknown salty springs on the lake bottom. Supposed years in which circulation did not take place in Lake Kinneret were conducive presumably also to a saline stratification.

The extreme scarcity of submerged vegetation has been noted by several authors. Waisel (1967) found submerged meadows of *Myriophyllum spicatum* and *Najas marina* in the somewhat protected Bay of Ginnosar as well as isolated stands of *Potamogeton pectinatus*. Por (1968a) emphasized especially the lack of Charophyta, otherwise fairly salt-tolerant, which are an important phytal substrate elsewhere in the area. The fauna of aquatic insects in the lake is very much impoverished, most probably because of the lack of both floating and submerged vegetation.

The benthos of the well-aerated bottoms contains several euryhaline and even halmyric organisms. Such are the already mentioned gammarid *Echinogammarus veneris* and also the oniscoid isopod *Halophiloscia couchii* which haunts the wet supralittoral of the Lake, as well as that of the Mediterranean sea shore. Among the ostracods, the dominant species is *Cyprideis torosa* a well-known euryhaline organism, together with the clearly marine *Loxoconcha galilea* and the African *Darwinula* cf. *stevensoni* (Lerner-Seggev, 1968). Among the harpacticoid copepods, there are no species of the characteristic freshwater family, the Canthocamptidae. Instead, there are species of clearly marine origin like *Pseudobradya barroisi* and *Nannopus palustris*, accompanied by the estuarine and euryhaline *Onychocamptus mohammed*, *Leptocaris brevicornis* and by three species of *Nitocra*. Curiously, among the cyclopoid copepods there are no marine species. Calanoid copepods are absent from the lake (with the exception of occasional specimens of *Arctodiaptomus similis* (Dimentman and Por, 1985). The sponge *Cortispongilla barroisi*, a dominant feature of the littoral, is considered by Racek (1974) to be a thallassoid species. The fact that it is unable to form gemmulae (resistant bodies), indicates a permanent presence of this species in the Rift Valley of the Jordan, at least since the last marine intrusion in the Pliocene.

The mollusks of the lake, though probably all euryhaline to some extent, are of freshwater origin. Perhaps the only exception is a recently found hydrobiid of the genus *Semisalsa* (Dimentman, unpublished). *Theodoxus jordani* is also common.

The history of the euryhaline genus *Theodoxus* goes back possibly to the Cretaceous times of the Gondwanian plate (Roth, 1987).

The planktonic world of the lake is peculiar. First of all, Lake Kinneret is dominated by a rather unusual combination of Dinoflagellates and Cyanophyta (Pollingher, 1978; Serruya and Pollingher, 1983). According to these authors the dominant dinoflagellate *Peridinium cinctum* forms blooms almost every year, a situation with no known parallel in other lakes. It seems that Lake Kinneret contained initially a predominant diatomacean phytoplankton; around 2,500 B.P., (Hellenistic and Roman periods) a predominance of chlorophytes started, probably due to incipient agricultural eutrophication; this was replaced about 1,600 B.P. by the present situation (Stiller et al., 1983–1984; Pollingher et al., 1984).

The zooplankton is typically tropical. Among the copepods the dominant and almost unique representative is *Mesocyclops ogunnus* an African cyclopoid identified for many years as the cosmopolitan *M. leuckarti* (Por, 1968b; Van der Velde, 1983). Among the cladocerans, the euryhaline *Bosmina longirostris* and the tropical species *Ceriodaphnia rigaudi* have to be mentioned. *Daphnia lumholtzi*, another tropical species disappeared from the lake in the last 3–4 decades, for unknown suspected anthropic reasons.

The fish fauna contains first of all the euryhaline cyprinodont *Aphanius mento* and the blenny *Salaria fluviatilis*, both resulting from earlier marine connections. (See for more data, Krupp and Schneider, 1988.)

There are several species of cyprinids which belong to the Mesopotamian fish fauna, to be mentioned below.

Interesting for our discussion at this point is the large diversity of the cichlid fishes, with a total of 7 species and subspecies in Lake Kinneret. Among them, *Tilapia zillii*, *Sarotherodon galilaeus* and *S. aureus* are known also from the rivers of the Mediterranean coastal plain and from Africa. *Astatotilapia flavijosephi*, *Tristramella sacra* and *T. simonis simonis* are endemic species whereas *T. simonis intermedia* was also shared with Lake Hula. The list is completed by two ancient tropical freshwater fishes, namely *Clarias gariepinus* and *Garra rufa* (Goren, 1974; Ben-Tuvia, 1978).

In all, the fish fauna of the Kinneret numbers 19 species. Ten species are secondary freshwater- and euryhaline species, almost all with Ethiopian affinities. Not only for tectonic reasons, but also because of the richness in cichlid fishes, Lake Kinneret appears as a small-scale African lake.

6.7. The Mesopotamian primary freshwater fauna

The primary freshwater fauna (Myers, 1938), in the broader sense given to this category by Banarescu (1970), comprises fishes, peracarid and decapod crustaceans, mollusks, oligochaetes, hirudineans and probably also planarians.

A certain care should be taken with taxa of smaller animals, which can be subject to passive transport. For the same reason, the anostracans, notostracans and conchostracans as well as the diaptomid copepods, considered by Banarescu as primary freshwater animals, have to be taken off the list. The importance of the primary freshwater animals is primary also for any discussion about the biogeography of the continental waters.

The present distribution of aquatic insects or micro-crustaceans which under certain circumstances can be dispersed overland depends more on contemporaneous environmental parameters and less on past geographic connections.

The inland waters of the Levant represent the southeastern branch of the Euphrates-Tigris system: they are the limnologically less stable half of the Fertile Crescent.

As shown by us (Por and Dimentman, 1985), the Mesopotamian Basin, where during the later Miocene the gradual closing of the Tethys took place, functioned as an important center for the evolution of euryhaline fauna. This fauna radiated from there, both in the direction of the Paratethys and of the Palaeomediterranean. In our present context, I want to broaden the importance of the Mesopotamian center: it was also without doubt an important craddle for the inland water fauna. We believe that new data from the largely uninvestigated basin of the Euphrates and Tigris will further emphasize the role of this Mesopotamian center.

The receding channel of the Tethys gradually transformed into river valleys and swamps. The Proto-Euphrates collected freshwaters from all over the Levant. Before the Pliocenic orogenesis it also maintained contacts with the Black Sea and Caspian fluviatile drainages. Some indications for these old contacts can still be found, for instance in the distribution of the bivalve *Dreissena* or of the cyprinid fish genera *Chondrostoma* and *Barbus*.

For a short time, the Mesopotamian freshwaters functioned even as a kind of "Freshwater Tethys", where African and Indomalayan species could freely intermingle or even have a common craddle. Several distribution patterns serve as an indication for this: Such is the present area of distribution of the fish family Mastacembellidae and of the genus *Garra* (Krupp, 1982). The prawn genus *Caridina*, the bivalve *Corbicula fluminalis* and the gastropods *Ferissia* aff. *wautieri* and perhaps *Melanoides tuberculata* represent, according to Kinzelbach (1987), the same pattern. The isolated population of a cichlid fish, *Iranocichla hormuzensis* in an Iranian tributary of the Straits of Hormuz (Coad, 1987) also indicates some very old African connections. At the latest, during the early Pliocene, desertification, rifting and orogenesis interrupted these tropical connections, too. Kinzelbach and Krupp (in Kinzelbach et al., 1987) call this distributional pattern "Palaeotropic" or "Oriental".

The fairly isolated Mesopotamian aquatic area of today presented a problem to the biogeographers. Berg (1948–49) considered it as part of the Palaearctic region. Banarescu (1970) saw in it a west-Asiatic sub-region of the Oriental region. In a later publication Banarescu (1977) considers the western Asiatic fish fauna to be a complex biogeographic crossroads: species from four regions are meeting there, but the degree of endemism is very high. There are Palaearctic genera like *Leuciscus* and *Chondrostoma* and genera which are more Mediterranean, like *Phoxinellus* and *Acanthobrama*; the Indoasiatic fauna is represented by *Tor canis, Mastacembelus simach, Clarias gariepinus* and the genus *Garra*. Although some Ethiopian affinities can also be found in the above-mentioned species, Banarescu considers only the Cichlidae as a real African element. Among the west Asiatic endemics, or taxa which have their main distribution here, Banarescu (op. cit.) mentions *Capoeta, Hemigrammocapoeta, Barbus* and *Alburnus* (Fig. 6.7).

Krupp (1987) disagrees on many accounts with Banarescu, especially regarding the taxonomy of the Cobitinae (loaches). He considers the Mesopotamian subregion as part of the "Indoasiatic" region (synonymous with "Oriental"). Kinzelbach (1987) speaks of a faunal mixture of "Paleoeuropean" and of south Asian origins.

The best solution would be to see the Euphrates-Tigris basin and its past and present tributaries as an independent Mesopotamian subregion in which, as it is always the case, a certain percentage of endemic taxa coexists with taxa originating from other biogeographic regions. The peculiar situation of the Mesopotamian subregion as very important crossroads during the Neogene, further justifies the

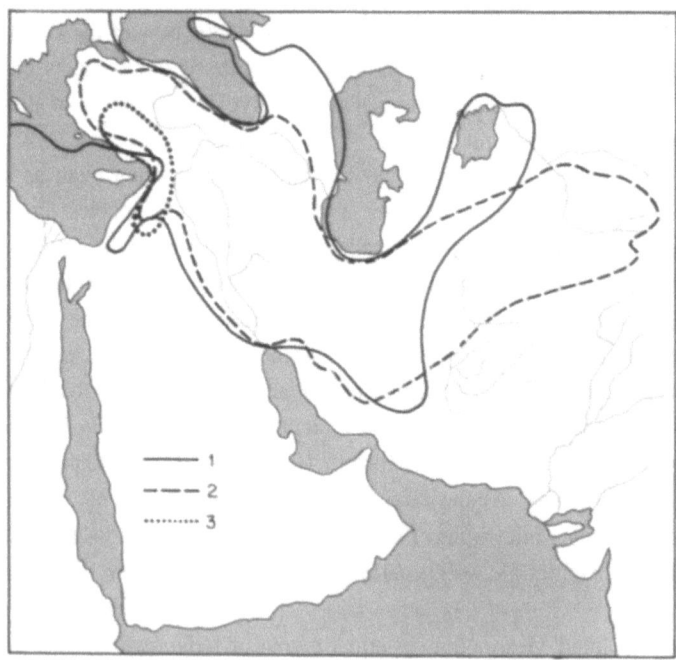

Fig. 6.7. Distribution of typical Mesopotamian fish genera: 1. *Barbus;* 2. *Capoeta*; 3. *Hemigrammocapoeta* (after Banarescu 1977).

present suggestion. Today the Mesopotamian subregion is one of the most isolated major freshwater areas of the globe.

Of the three barriers which isolate the Mesopotamian rivers from the other basins, i.e. the alpine mountains in the north, the Syrian and Sinai deserts in the south and the Iranian and Sindh deserts in the east, the last one remained the less impermeable one. But it is still a very difficult hurdle to pass.

In our discussion of this oldest primary freshwater stock of the Levant we shall have to keep in mind that, though mainly Oriental in its connections, the Mesopotamian sub-region also presents strong Pontocaspian, i.e. Palaearctic relations. In this sense we are close to Banarescu's (1977) basic views.

6.8. The separation of the river basins

As shown above, during the Miocene and for much of the Pliocene, the waters of the Levant drained to the east and to the Euphrates (see p. 15). A gradual separation followed, evidently isolating first of all the more distant Jordan basin in the south. As shown by Kinzelbach (1987), the Mesopotamian species living in the Jordan valley had old and direct connections with the Euphrates and hardly received any later additions through the Orontes. If there is any similarity between the fauna of the Jordan and that of the Orontes, this dates back to Pliocene times, or occurred to a limited degree via the Damascus basin. Like Por (1975b), Kinzelbach was first in favor of a sweep-stake connection between the major Levantine river basins. However, this author now authoritatively proved that the Lebanese mountains and the Litani basin itself did not serve as an intermediary between the Orontes and the Jordan (Kinzelbach,1987). The ancient "direct" eastward connection with the Euphrates left a few relictic aquatic environments behind in the oases of the Syrian and Jordanian desert.

The oasis of Damascus received its population of *Caridina fossarum syriaca, Unio crassus damascensis, Corbicula fluminalis* and *Potamon potamios setiger* according to Kinzelbach (1987) from the Euphrates through the oasis of Palmyra. Of them, *Corbicula* may also have reached the Jordan by this way. Indicating a close connection, several species of fishes are shared by the oasis of Damascus and the Jordan; these are *Capoeta damascina, Hemigrammocapoeta nana, Nemacheilus insignis* and *Tristramella sacra* (Krupp, 1987). Kinzelbach (op. cit.) mentions the finding of a fossil population of the Mesopotamian *Melanopsis nodosa* in the Pleistocenic Ubeidiya formations of the Jordan (Tchernov, 1973) as proof for old direct connections between the Jordan and the Euphrates: this species has not been reported yet from the Orontes.

In general there are few species which are common to the Orontes and the Jordan. *Capoeta damascina* is an example, in addition to *Corbicula* and *Melanoides*, which have been mentioned above.

The recent data by Kinzelbach and Koster (1985) indicate a beginning of evolutionary divergence between the populations of the prawn *Atyaephyra desmarestii orientalis* which inhabit both the Jordan and the Orontes basin. They are isolated by the basin of the Litani, where this prawn is not found. As mentioned elsewhere, *Atyaephyra* is found both in the waters of the oases of El Azraq and of Damascus (Fig. 6.8).

The Mesopotamian biota of the Jordan valley are as a rule different from those of the Orontes, either at species or at subspecies levels (Kinzelbach and Roth, 1984;

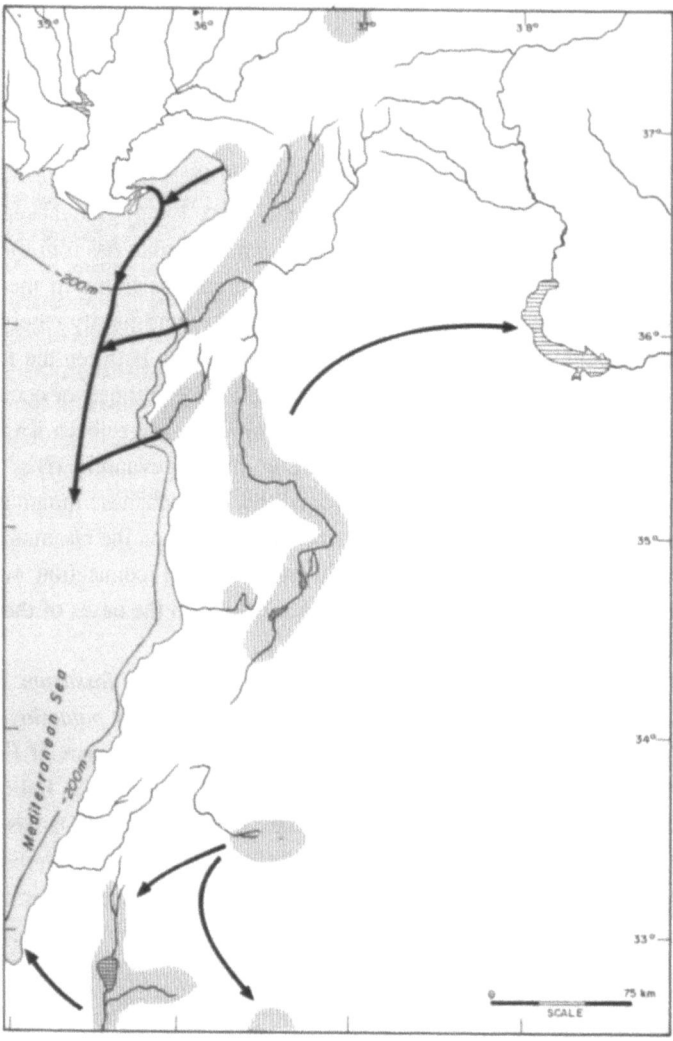

Fig. 6.8. Distribution of the freshwater prawn *Atyaephyra desmarestii* in the Levant. Presumed pathways of expansion are indicated by arrows (after Kinzelbach, 1987).

Kinzelbach, 1987): *Potamon potamios palaestinensis* vs. *P.p. ghab*; *Unio terminalis terminalis* vs. *Unio t. delicatus*; *Leguminaia saulcyi* vs. *L. wheatleyi*.

Among the fishes the differences are more considerable. According to a recent tabulation by Krupp (1987), *Barbus canis* and *B. longiceps* live in the Jordan section of the Rift Valley and *B. chantrei, B. grypus, B. luteus* and *B. pectoralis* in the Orontes section. Similarly *Hemigrammocapoeta nana* lives in the Jordan and *H. culiciphaga* in the Orontes. In the Jordan graben three endemic species of *Acanthobrama* are found, while in the Orontes basin only one is known. The endemic *A. tricolor* of the Damascus basin possibly connects between the Orontes species and its congeners from the Jordan. On the negative side, none of the three species of *Leuciscus* of the Orontes reached the Jordan basin.

Krupp (op. cit.) summarizes the ichthyogeographic picture as follows: Out of its 18 primary freshwater species, the Jordan shares 7 species with the Orontes; 3 species are shared with the Euphrates and 6 have their closest sibling in Mesopotamia. By comparison, the Orontes shares 18 of its 32 species of primary freshwater fishes with the Euphrates.

In the northern section of the Levantine Rift Valley connections with the Euphrates continued probably into sub-recent times. Por (1975b) hypothesized a "headwater capture" relation between the Orontes and the Euphrates. It is the merit of Kinzelbach (1980, 1987) that the detailed pathway of this connection has been put into evidence. Accordingly the presently endorheic river Qwaik in the oasis of Aleppo was a key element connecting the middle course of the Orontes with the Euphrates. Kinzelbach also gives a list of species which presumably used this waterway.

Further possible connections were through the intermediary of the rivers Afrin, Kara Su and Ceyhan. The first two are presently connected with the Orontes through lake Amiq. River Ceyhan might have been confluent with the lower Orontes during a low eustatic sea level (see Fig. 6.8).

In the south, the biogeographic history of the El Azraq oasis is somewhat unclear. As mentioned above, the phreatic sheath under Gebel Druz emerges there, forming two strong freshwater springs. These form small permanent pools which overflow during the rainy season to form a large saline lake and several more pools. The fauna of the oasis has been briefly presented by Scates (1968).

The presence of *Atyaephyra* in El Azraq (Kinzelbach, 1987) indicates old connections with the Jordan; that of *Aphanius sirhani* (see above p. 120) may suggest an old estuarine connection either to the east or to the west. Schütt (1983) lists a fairly rich molluscan fauna from El Azraq, which contains *Theodoxus jordani, Semisalsa contempta, Melanoides tuberculata, Melanopsis praemorsa costata, Radix peregra* and, of course, the amphibious *Bulinus truncatus*. The thought that El Azraq might represent a very old connection between the Jordan graben and the rivers of the nascent Persian Gulf cannot be easily discarded.

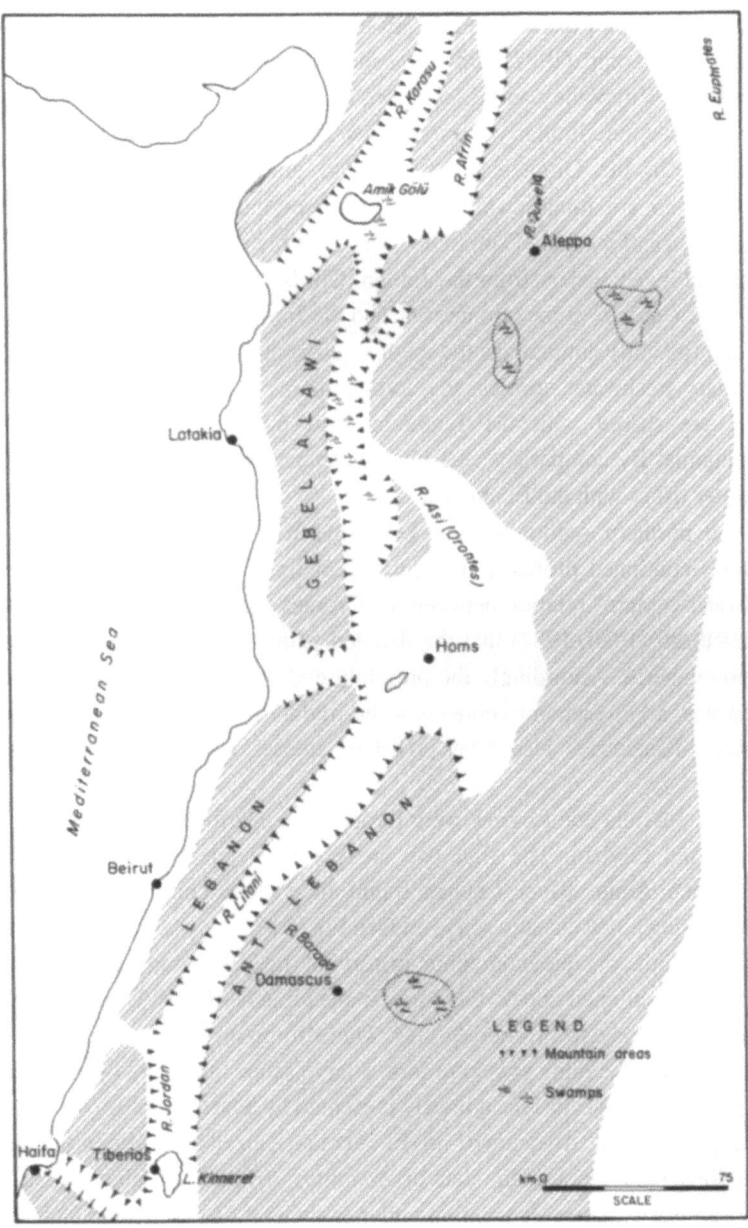

Fig. 6.9. The northern part of the Levantine Rift Valley with the Orontes and Litani Basins (after Weulersse, 1940).

6.9. River Orontes and its complex history

According to Kinzelbach (1987) the Orontes (Nahr Asi) as we know it today, is a patchwork joined together only 6,000 years ago by the unification of three separate rivers. We shall not go into the geological details of this event here, nor discuss the exact timing of it. We shall briefly describe the three basins and the tribulations they underwent in historical times.

The Lower Orontes starts in the Lebanon, north of the Baalbek water divide with the Litani (Fig. 6.9).

Near Hermel, at Ain el Zarqa, in a stretch of about 500 m of the river bed, more than 12 cubic meters of water per second gush out into the river bed.

After a short and very steep gorge, the Orontes emerges in the plain of Lake of Homs. This lake was temporarily a terminal swamp lake, confined to the north by the basaltic sill of Rastan. Following de Vaumas (1957a, b), Kinzelbach (1980) considers that the Lower Orontes discharged at least temporarily into the Mediterranean by the intermediary of Nahr El Kabir south. This coastal river of today shares several species with the Orontes, unlike other coastal streams of Syria: such are the bivalves *Potomida littoralis semirugata, Unio terminalis delicatus, Pseudunio homsensis, Leguminaia wheatleyi* and *Corbicula fluminalis* as well as *Potamon potamios setiger* (Kinzelbach, 1987). There are also several fishes which

Fig. 6.10. "Naura's", giant irrigation wheels on the Orontes near Hama (old photograph from Weulerrse, 1940).

reached this Mediterranean river through the old connection with the Homs basin: *Phoxinellus zeregi, Leuciscus cephalus* and *Hemigrammacapoeta culiciphaga* (Krupp, 1987). An old contact with the Litani is seen today as less probable.

The Lake of Homs (Lake Qatina) of today is a semi-artificial lake built in the 2nd millenium B.C. In 1938, the dam was raised by a further 2 m and the surface of the lake increased from 43 to 60 square km (Dubertret and Weulersse, 1940). Another dam formed Lake arRastan, a few tens of kilometers downriver from Hama. The famous "Noria's" or more correctly "Naura's", the old irrigation wheels of Hama, are still active today, and represent real monuments of classsical hydraulic technology (Fig. 6.10). Their average diameter is of 10 to 12 meters, but one of them is known to measure 22 meters. These irrigation giants were moved by the stream and could thus distribute water to a considerable height above the water level. This stretch of the Orontes is presently suffering from heavy pollution and at ar Rastan oxygen levels are often very low (Kinzelbach, 1980; Moubayed, 1986).

Another gorge is cut by the Orontes at Cheizar and the river emerges into the vast swampy plain of the Ghab. According to Weulersse (1940), "the Ghab is for the Orontes what the Dead Sea is for the Jordan; but here the climate is more humid and the elevation higher (about +170m); the Orontes can therefore escape(!)".

The Middle Orontes is without doubt the most important section of this river. It corresponds to the geologically very recent Ghab depression, probably also an active sector of the Levantine Rift Valley. The early connections of the Middle Orontes with the Euphrates have been discussed above. The Middle Orontes, too, ended for some periods in the swamps of the Ghab and Aharne, which correspond probably to old terminal lakes. It also had a connection with the coastal rivers Nahr Marqiya and Nahr el Kebir north. The first of these was colonized by the eminently successful *Capoeta damascina* and the second also by *Leuciscus spurius* (Krupp, 1987). *Potamon potamios ghab* likewise reached the Mediterranean shore through these connections. Later, the Middle Orontes lost its coastal drainages. At a certain stage, finally the river cut through the sill of Karkor and obtained an opening to the sea through its present lower section.

At the beginning of this century the swamps of Ghab covered 300 sq. km and the swamps of Aharne 130 sq. km (Dubertret and Weulersse, 1940). Starting in 1954 the swamps of the Ghab-Aharne-Roudj system were gradually drained and transformed into lands for irrigational agriculture. Storage dams were established, like those at Maharda and Aharne.

As shown by Krupp (1980), the network of irrigational channels which replaced the old swamps, some of them with concrete walls and bottoms, led to a catastrophic impoverishment of the original fish fauna. Salinities increased in the channels and a chlorinity of 0.5 ppt was measured in one of them. The backwaters, storage lakes and channels left over have a very reduced fish population. Important commercial fish populations, like that of *Clarias gariepinus* disappeared from the Ghab basin.

The Middle Orontes with its many endemic and rare fish species is presently suffering another ecological damage: Krupp (1980) mentions that several species of fish were introduced to a number of storage lakes of the area. Such are *Mastacembelus simach* and *Silurus glanis* brought from the Euphrates, *Tilapia* spp. and *Cyprinus carpio* brought probably from the Lake Kinneret area.

After crossing the basaltic sill of Karkor, the Orontes emerges into its lower stream bed. It skirts the southern edge of the valley of Lake Amiq where it captured the emissary river of this lake, River Kuchuk Asi or Ak Su. To reach the sea, the Lower Orontes turns sharply to the west and cuts through a final basaltic sill at

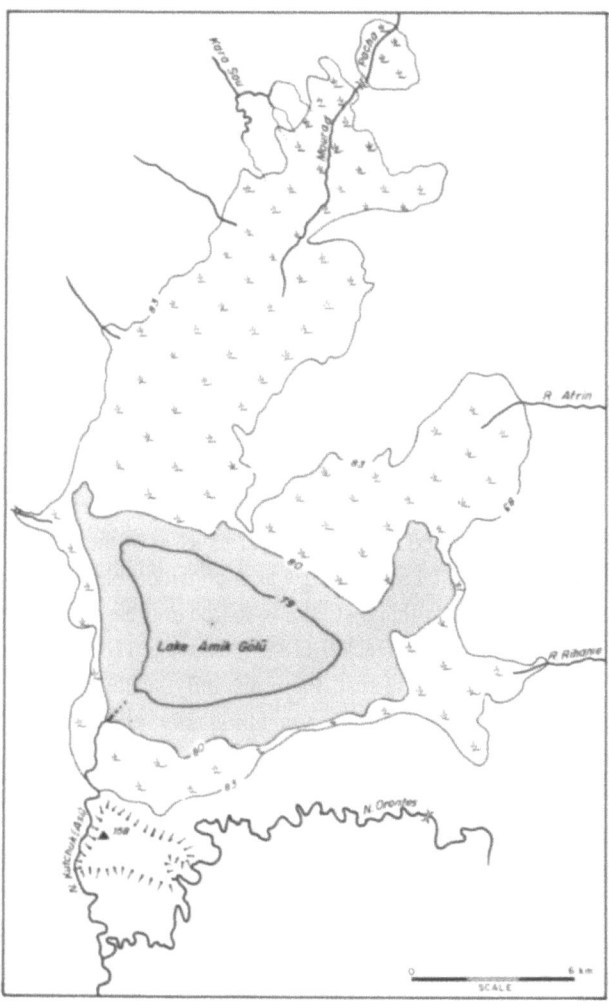

Fig. 6.11. Lake Amiq before its drainage (after Weulersse, 1940).

Antakiya before descending to the Mediterranean. The wide Amiq valley which is now drained by the Orontes contained Lake Amiq (Amiq Gol, Lake of Antiochia, Lake of Antakiya) (Fig. 6.11).

Before its drainage around the middle of this century, Lake Amiq was surrounded by an extensive marsh into which two important rivers flew: the Afrin and the Kara Su. The valley of the latter represents the northernmost section of the Rift Valley (Mazloum, 1939). Both rivers start on the southern slopes of the Amanus Mountains. The Kara Su is 90 km long and the Afrin 149 km long. The first has almost the double of the discharge of the second. According to Mazloum (1939) the swamps occupied an area of 22,000 hectares and the open lake some 9,000 hectares. According to Dubertret and Weulersse (1940) the swamp covered 600 sq. km and the open water 90 sq. km. (Fig. 6.11). River Kuchuk Asi, the short lake emissary discharges after only 15 km into the Orontes. The declivity along the last stretch of the Orontes to the sea is very great: 79 m of level difference, for 56 km of river bed. The contribution of the Lower Orontes to the combined outflow is 30 cubic meters/second (Kinzelbach, 1980).

During most of the Pleistocene, river Kuchuk Asi drained the lake to the Mediterranean (Gruvel, 1931). The capture of the Kuchuk Asi could be seen as an almost sub-recent event.

As shown by Kinzelbach (1987), the basin of Lake Amiq has been an important meeting point of aquatic faunas: through fluviatile sweep-stakes river Afrin was connected with the Upper Euphrates; on the other hand, the Afrin, or the emissary river Kuchuk Asi at times connected with the Anatolian river Ceyhan. Thus a pathway for the Palaearctic fauna was also in existence. After the capture of the Amiq by the Lower Orontes, a third fluviatile system was linked to this aquatic melting pot. Closeness to the Mediterranean also contributed a catadromic fish fauna, of which the eels formed the major fisheries asset of the lake.

The complete drainage of Lake Amiq was a considerable loss for science. The hydrobiological knowledge of this lost lake is scarce and based only on authors like Locard (1883) for mollusks and Heckel (1843), Sauvage (1882), Lortet (1883) and Gruvel (1931) for fishes. For the rest, there are only scattered data. According to Aharoni (1942) the darter bird *Anhinga melanogaster chantrei* lived in the lake, together with an extremely rich aquatic avifauna.

The unclear picture which emerges is that of a swamp-surrounded lake, the faunal diversity of which was matched probably only by that of Lake Hula in northern Israel, the second-largest freshwater lake of the Levant (see below). However, the dimensions of the total faunal loss will never be known.

Before leaving the Orontes, another beautiful quotation from Weulersse should be added: "The Orontes appears thus as a strangely composite river; as a matter of fact it is a succession of 5 different streams: a Lebanese mountain torrent till Homs, a Syrian highland river around Hama, a rift drainage in the Ghab and another one in the Amiq and finally a Mediterranean coastal river after Antakiya." This is very reminescent of the division of the Jordan into the hydrologically different Upper, Middle and Lower sections.

Krupp (1987) considers that river Qwaik already belongs to the Euphrates province and not to the Levant. The importance of the rich fish fauna of this endorheic basin has already been emphasized above. During the earlier decades of this century, pollution and irrigational projects severely reduced the faunal inventory of the Qwaik. Krupp (1980) found there only *Phoxinellus zeregi, Garra variabilis, Alburnus sellal* and *Chondrostoma regium*.

6.10. The Ethiopian connection

The influx of Ethiopian (= Afrotropical) species into the fauna of the Levant has been extensively discussed (Por, 1975b, 1987; Dumont, 1979). Several circumstances have to be taken into consideration when one discusses the African influence in the inland water fauna of the Levant. First and foremost, there has never been a continuous nor even contiguous riverine connection between the Levant and Africa, at least since pre-Messinian times. No river is known to have crossed the desertic expanses to the Nile, neither had the Jordan graben any aquatic connection to the Red Sea. These barriers were never lifted.

Ethiopian faunal elements can be found therefore in restricted ecological groups in which euryhaline adaptations and/or a good capacity for active or passive spreading exists. Furthermore, whatever Ethiopian influence is felt in the inland waterfauna of the Levant, this is mostly limited to the south, i.e. to the Jordan valley and the coastal plain of Israel. There are only isolated cases of Ethiopian species extending into Lebanon and Syria.

Indeed, as discussed above (p. 128), there is an older Paleotropical stratum of species which lived in the Early to Middle Miocenic peri-Tethyan freshwaters. Here we are referring exclusively to presence of African species which either did not reach the Mesopotamian waters in the distant past or are of a too recent origin to have done so.

Tchernov (1988) considers that the aquatic fauna found in the Jordan valley harbored for some time Gondwanian species cut off from the African mainland through the Saharan desertification. The "Negev fauna" of the late Burdigalian (18–19 million years ago) contained the fish genus *Lates* as well as the soft-shelled freshwater turtle *Trionyx* and also *Crocodylus*. During the Pliocene and the early Pleistocene, cichlids were common in the Jordan valley. The Early Pleistocenic Ubediya formation, mentioned above, also contains the turtle *Trionyx triunguis* and *Crocodylus niloticus*. According to Tchernov (1988) *Hippopotamus amphibius* was present in the Jordan valley till Middle Pleistocene. Interestingly however, the mollusk fauna which accompanied this African aquatic macro-fauna was composed of non-African species (Tchernov, 1973, 1975b, 1987). We have the definite impression that during the Plio-Pleistocene only strong-swimming animals could reach the Jordan valley.

In an actualistic way, this hypothesis is strengthened by the ongoing forays of the Ethiopian aquatic fauna into the coastal streams of Israel. *Hippopotamus* survived there until the Bronze age (Haas, 1953) and *Crocodylus* till 1912 (Yom-Tov and Mendelssohn, 1988). *Lates macrophthalmus*, the Nilotic fish, was found in archeological sites of the coastal plain (Lernau, 1988). *Clarias gariepinus* and *Trionyx triunguis* live in the coastal streams today. The nilotic fish *Tilapia zillii* is present in the streams, even north of the Litani in Lebanon, and frequently reported from the open sea (Chervinski, 1987). Even some Nilotic mollusks, like *Cleopatra bulimoides, Pila ovata* and *Lanistes sp.* are occasionally reported from the Mediterranean coast of Israel (Barash, in lett.). Kinzelbach (1987) mentions a population of *Cleopatra* in Nahr al-Kabir south, living there in historical times. Floating water hyacinths *Eichhornia crasssipes* from the Nile often reach the Mediterranean shore of Israel. According to Ortal (in lett.), they often carry along viable specimens of waterbugs and waterbeetles which belong to Nilotic species.

Everything points towards the influence of the Nile waters carried eastward by the anticlockwise longshore current. This is an important and perhaps the only real aquatic connection with the aquatic fauna of Africa. At times when the flow of the Nile was one order of magnitude greater, about 12,000 years ago (see above p. 31), the Nilotic flood could easily carry animals able to survive for some time the impact of brackish water. Active swimmers had an even easier task. Together with this, we have to consider the low sea-level situations when the distance between the southernmost estuaries of the Israeli coastal plain were much closer. Real tributary connections were probably established only by the western Nile branches and Wadi El-Arish. Even though periods of low seawater level were usually not the periods of increased freshwater inflow, intermediate and transitional situations were many (see p. 38).

The Jordan valley was therefore intermittently colonized through the intermediary of the coastal streams by actively swimming aquatic vertebrates. Only some of them, namely the cichlids *Astatotilapia flaviijosephi, Tristramella simonis* and *T. sacra* evolved into endemic species. Probably the principal "entrance gate" to the Jordan valley was the Yezreel Valley, often flooded and even recently presenting swampy connections between the headwaters of Nahal Qishon and the streams of the Bet Shean valley (Por, 1975b). As mentioned by Krupp (1987) the Qishon valley harbored *Acanthobrama lissneri* and *Garra rufa*, two fishes also found in the Jordan valley.

The siluroid *Clarias gariepinus* already mentioned also reached the Orontes, but this is a fish with considerable capacities for overland movement and air breathing. Several populations of *Clarias* live also in isolated Sahara oases and are considered to be well-adapted desert fishes (Dumont, 1979). Similar is the case of the gastropod *Bulinus truncatus*, which is sufficiently drought-resistant and amphibious to be the only widespread "African" gastropod in the Levant (Kinzelbach, 1987).

Other Ethiopian species are found among the inland water cladocerans and copepods. The Ethiopian origin of *Bosmina longirostris* and of *Daphnia lumholtzi* are clearly shown by Dumont (1979). To these one has to add the *Ceriodaphnia*. The first two cladocerans are found also in the basin of the Euphrates. The Nilotic *Afrocyclops gibsoni* was reported from Lake Hula as well as from the Kinneret and the Lower Jordan.

Mesocyclops ogunnus, the dominant copepod of Lake Kinneret and a frequent inhabitant of small still waters and slow-flowing streams in Israel is also an Ethiopian species. Dumont (1979) adds two cases of Saharan species, which he identified in the waters of Transjordan: *Cryptocyclops linjanticus* and *Metadiaptomus chevreuxi* (the latter also in the Euphrates basin). In this category of desert copepods of African origin one also has to mention the calanoid *Neolovenula alluaudi*, an Ethiopian species, present in the Nile Delta in the Negev and the Syrian desert, but also circum-Mediterranean (Dumont, op. cit.; Dimentman and Por, 1985; see also below, p. 156).

Possibly the ostracod *Darwinula africana* identified by Martens (in litt.) in the Dead Sea springs belongs to the same category of desert-water crustaceans. Unfortunately little is known about the inland water ostracod fauna of the Levant.

In the case of the above-mentioned small crustaceans, passive transport should be considered. They are either able to form resting forms or reproduce in small water accumulations. Distribution by floods or by winged animals results in a kind of "island hopping" through the desert barriers.

Aquatic insects are even better able to do this. Wewalka (1986), analyzing the distribution of 72 species of dytiscid Coleoptera of the Levant, recognizes a whole group of "eremic" species, some of them also able to withstand saline waters. Of these, *Potamonectes cerisyi*, *P. lanceolatus* and *Eretes sticticus* widespread in Sinai and in the Negev are worthy of mention. Wewalka (op. cit.) characterizes 6 dytiscid species as Ethiopian, among them *Canthydrus diophthalmus*, which advanced north to the Orontes basin. A somewhat similar picture is also given for the hydrophilid Coleoptera by Jäch (in lett.).

Without doubt, the Ethiopian connection is the strongest among the skillful flyers which are the Odonata. Dumont (1979; in press) identifies a considerable number of Ethiopians among the dragonfly and damselfly fauna of the Levant. Some of them are clearly "Saharan" like *Orthetrum taeniolatum*, *Sympetrum decoloratum* and *Paragomphus sinaiticus*; others advanced along the waterway of the Nile, like *Ischnura senegalensis* and *Orthetrum ransonetti*. Two species of tropical Africa have isolated subspecies in the Levant, namely *Urothemis edwardsi hulae* and *Rhyothemis semihyalina syriacum*. Finally, other species probably used the mountain connection along the Arabian Red Sea coast in order to reach the Levant from the Ethiopian enclave of Yemen. This pattern is extensively discussed by Schneider (1987) for the distribution of the genus *Pseudagrion*. A further example of this distributional pattern along the Red Sea shore is that of *Crocothemis sanguinolenta* (Schneider, op.cit.). Although this author admits that *Pseudagrion torridum* with its Levantine subspecies *P. torridum hulae* advanced along the Nile, he states that "most Middle Eastern Odonata with African affinities invaded the Levant along the Red Sea coast of Arabia and not via the Nile and Sinai". Rich and new data on the distribution of the Odonata in the Levant are given by Krupp and Schneider (1988).

According to Botosaneanu and Gasith (1971) and Botosaneanu (1973, 1974) there are several Ethiopian species among the Trichoptera of the southern Levant. Botosaneanu, too, tends to see a connection with the Ethiopian faunal enclave of southwestern Arabia, rather than with the Nile. Interestingly, he also mentions the capacity of some trichopteran species to survive in temporary waters by retreating into the protected subterranean domain.

Ethiopian species are fairly common among the Chironomidae of Israel, especially in the Jordan valley and Lake Kinneret (Kugler and Chen, 1968; Kugler and Reiss, 1973). Five species of this Ethiopian stock reach Lebanon from where they were reported by Moubayed and Laville (1983) and Moubayed (1986).

Ethiopian species in the Levant are also found among the Hemiptera (Linnavuori, 1960, 1964). Among them, we would like to mention the Belostomatidae, several Veliadae, genus *Micronecta* and the obvious *Ranatra parvipes vicina*.

Moubayed (1986) gives a summing-up of the Ethiopian aquatic entomofauna in the Levant, in an Appendix to his thesis. Most interesting in this list are the representatives of the ephemeropteran genera *Afronurus* and *Oligoneuriopsis*: may flies are not known to be very strong on wing!

The leech *Limnatis nilotica*, common in southern Israel and recently reported also from Lebanon (Dia, 1983; Bromley, in press), was probably transported by prey animals.

The only mollusks from the Levant known to have expanded to the south and populated the Nile valley are *Unio tigridis* and *U. mancus* (Kinzelbach, 1987). Like other unionids, these shells have glochidia and therefore passive transport by host fishes cannot be ruled out.

In general, there are few cases in which the Levantine province served as a passageway for aquatic biota into Africa. Again passive transport by birds could be responsible for the forays of Palearctic microcrustaceans like *Moina, Daphnia* or *Arctodiaptomus* to Africa.

Insects present many instances of "circum-Mediterranean" distribution. This pattern poses a difficult problem to biogeographers (see Por, 1975b; Por, 1987). Flying organisms probably had ample opportunity to cross Mediterranean straits both ways, especially during low sea levels. Besides, as repeatedly suggested by Por (op. cit.), human influence by either navigation or agricultural exchange probably enhanced distribution of adventive organisms around the Mediterranean.

Entomologists are inclined to characterize some species of aquatic insects as "Eremic" (e.g. Hemiptera, Coleoptera). Among the Odonata and the Copepoda Diaptomidae, Dumont (1979) mentions also "Saharan" species. This is a fauna typical for the "Palaeoeremic" biogeographic area, proposed by Por (1975b) for the terrestrial fauna.

6.11. The Palearctic influx and its limitations

De Lattin (1967) is of the opinion that what we call Palearctic primary freshwater animals are descendants of species which retreated during the Glacials into several refugia. The important refugium in western Eurasia was, in his opinion, the Pontic refugium. From there, fishes and other freshwater animals repeatedly expanded over Europe, after the Terminations. Most of the primary freshwater fishes, mollusks and decapod crustaceans of the Levant originated in the Mesopotamian river systems, which as shown above (p. 127) had connections with the rivers of the Black Sea basin. In this sense, among the primary freshwater species found in the Levant there are probably several species which spread both to the north and to the south, from the common Mesopotamian center. Care should be taken before calling them species of Palearctic origin.

However, when one deals with the aquatic fauna of the phreatic waters, of the springs and of the brooks, De Lattin's view cannot be sustained. This basically cold-stenothermic fauna did not necessarily suffer when the rivers and lakes froze over. As seen below, this fauna of insects or permanently aquatic stream-haunting animals could easily expand their range from stream to stream. Real Palearctic species are to be found among the stream-loving coldwater species and can be relatively easily separated from the Mesopotamian ones.

This Palearctic, Euro-Siberian freshwater fauna is therefore basically a cold-stenothermic mountain fauna of insects, benthic entomostracans, flatworms, oligochaetes, leeches, sphaeriid bivalves and genera of cyprinid and cobitid fishes. It will probably be well-represented among such taxa like the gastrotrichs, water-mites and tardigrades which have not beeen studied so far in the Levantine waters.

The southward advance of the Euro-Siberian aquatic fauna in the Levant depended chiefly on the availability of cold-stenothermic waterbodies. This is an important definition, because, for the primary-aquatic, fluvio-lacustrine fauna, there were no post-Pliocenic fluviatile connections between the Levant and the rivers of Europe and northern Asia. In a sense, the situation is similar to the lack of fluviatile contact with the Ethiopian realm (see above, p. 136). The difference is in the fact that near the limits of northern Levant there was always a wealth of coldwater streams.

Only those biota which were able to settle the springs and streams of the Anatolian mountains, of the Taurus, Zagros and Caucasus, i.e. of the mountain barrier which separates the Levant from Asia and eastern Europe could advance into our area. Once in the Levant, the chain of the Cis-rift and Trans-rift mountains (Por, 1975b) provided a pathway of spring and stream environments for these Palearctic species. The subdivisions along this pathway and its ultimate southward limits will be discussed below.

In this sense, an approximate separation of the Palearctic stenothermic fauna from the Mesopotamian riverine and eurythermic fauna is possible. Though typical Palearctic biota may also be encountered in the two large river systems of the Orontes and the Jordan, they are mainly found in the headwaters and the other mountain waters of the Levant.

The Palearctic element is especially evident in the basin of river Litani. Above we mentioned (p. 129) that the Litani is a basin which has probably never been settled by the Mesopotamian freshwater fauna. Its relatively poor fauna is therefore primarily composed of the stenothermic Palearctic species.

What is after all this Palearctic fauna of the Levant? Representatives of it are to be found in all of the major taxa; however, they are predominant in several of them. These are the Tricladida (Bromley, 1974); the Amphipoda of genera *Niphargus* and *Gammarus* (Alouf, 1982; Herbst and Dimentman, 1983; Karaman, 1986); Copepoda Harpacticoidea, family Canthocamptidae (Por, 1983c); Hirudinea Erpobdellidae (Ruckert, 1985; Bromley, in press); Gastropoda Lymneidae and Bivalvia Sphaeriidae (genus *Pisidium*) (Schutt, 1983); Ephemeroptera (Demoulin, 1973; Por, 1975b; Dia, 1983); Plecoptera (Zwick, 1972; Bromley, in press); Heteroptera (Nieser and Moubayed, 1985); Simuliidae (Crosskey, 1967; Moubayed and Clergue-Gazeau, 1985); Coleoptera Gyrinidae (Jäch, in press); Pisces Cobitidae (Krupp, 1987). According to Martinez-Ansemil and Giani (1987) the Levantine fauna of aquatic Oligochaeta is composed overwhelmingly of Palearctic species. This is a confirmation of Cernosvitov's (1938) previous statement, according to which the Middle East is one of the richest areas in oligochaetes, because "no important barrier separates it from the Euro-Siberian and east Siberian faunas".

Some representatives of this rich Palearctic oligochaete fauna are *Peloscolex kurenkovi* (Pascar-Gluzman and Dimentman, 1984) *Nais bretscheri, Psammoryctides barbatus* and *Stylodrilus lemani* (Moubayed et al., 1987).

As mentioned above most of these taxa have in common the fact that they tend to concentrate in running waters, preferably fast-flowing, well oxygenated and stenothermic, but most importantly, "completely" fresh. Some of them can probably survive well either in wet river mud or vegetation and retreat into subterranean waters of the springs or into the hyporheic environment (i.e. the accompanying subterranean flow) of the rivers.

The expansion of the Palearctic species along the mountain chain systems was possible either through the flight of the imagines, or in the case of the non-flying taxa, through torrential rains, headwater capture and tectonic changes. In at least some cases, hydraulic engineering helped, too. For instance, many tens of springs of the Judean mountains were connected into a huge system of aquaducts which supplied Hellenistic and Roman Jerusalem with drinking water. All the major classical cities in the area, like Askalon, Samaria, Caesarea and Acres, had their network of aquaducts.

The above-mentioned qanat system (p. 110) probably provided a free access into the cool subterranean waters for many stenothermic species.

6.12. The Lebanese rivers

Lebanon is a hinge-area between the southward flowing Jordan and the northward flowing Orontes (Nahr Asi) (Fig. 6.12). The valley of the Bekaa is a relatively

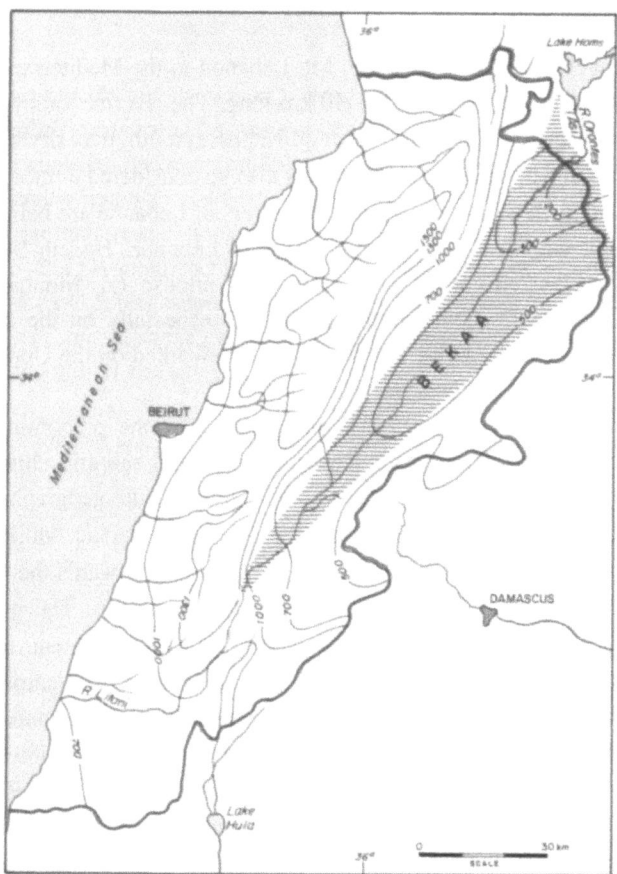

Fig. 6.12. The Basin of River Litani and of other Lebanese streams. The valley of Bekaa is shaded. Heavy line indicates political frontier of Lebanon. Average annual isohyetes in mm (after Moubayed, 1986).

narrow (4–10 km) and high-altitude valley (800–900 m) wedged in between the mountain chains of the Lebanon (from 600–1000 m in the south to 3000 m in the north) and the chain of the Antilebanon with average elevation of 2000 m, and highest altitudes in the south. Towards the north, inside Syria, the Bekaa opens up and descends to lower elevations.

Precipitations in the area are locally abundant, reaching an annual 1,400–1,500 mm in the northern Lebanon mountains. In the Bekaa, precipitation decreases from 800 mm in the south to 200 mm in the north, where the valley opens up to the Syrian desert.

River Litani is the main river of the Bekaa valley, running south. Near Beaufort castle, the Litani turns abruptly to the west, crosses the coastal mountain range and opens into the Mediterranean. South of the village of Hasbaya, a second river of the Bekaa, river Hasbani, flows parallel to the Litani with sometimes less than 5 km separating the two rivers. Ultimately, the Hasbani converges with rivers Dan and Banias to form the Jordan. North of the water divide of Baalbek at 1,100 m altitude, the Bekaa is drained by the Orontes. In general, the rivers of the Bekaa have barren or cultivated river banks.

The coastal rivers descending from Mt. Lebanon to the Mediterranean are short and steep: they are as a rule only 30–40 km long. The riverine valleys are narrow, deep and well-forested. Some of them, e.g. Nahr Beyrouth, may dry out during the height of the summer. The lower Litani may also be considered a littoral river.

Largely unknown till the last decade, the rivers of Lebanon are being intensively investigated by scientists from the University of Lebanon, Hadeth-Beyrouth (N.J. Alouf, A. Dia) and from the University of Toulouse (Z. Moubayed and his colleagues). Dia (1983) concentrated his studies especially on the coastal rivers Nahr Damour and Nahr Aouali. Moubayed (1986) worked on the Upper courses of the Orontes, the Litani and Nahr Beyrouth.

The hydrography of the region is complicated by the predominantly karstic nature of the substrate. This, combined with the general sub-arid climate, imposes an irregular and often discontinuous river flow, both in the geographic and in the seasonal sense. Of especial interest are the two large karstic "dolina's", where spring water and melting snow accumulate in the winter; towards the summer they nearly dry out, primarily because of the karstic percolation. The ponor lake of Yammoune used to reach winter-extensions of 10/5 km. At present, because most of the surface-waters are captured by irrigation channels, the temporary lake is reduced to a miniature fragment of its original extension (Moubayed, 1986). However, the majority of the waters of this mountain lake form a subterranean river which emerges after some 20 km in the impressive spring of Ein-Zarka, which at 600 m altitude is the main headwater of the Orontes. It is interesting that according to Moubayed (op.cit.), the surface waters of the Yammoune used to drain towards the Litani.

On the western bank of the Litani, north of Mansoura there is another dolina which used to contain the swampy area of Ammiq (not to be confused with Lake

Ammiq or Lake Antiochia on the lower Orontes). The swamp, probably a Pleistocenic lake, occupied an area of 150 square kms; this is today much reduced through land reclamation. During 2–3 months in the summer the swamp dries out completely (Moubayed, op. cit.).

The course of the Litani is interrupted by the artificial lake Qarun. This lake and the southern stretch of the Litani are heavily disturbed and polluted.

Water temperatures in the rivers of Lebanon are much lower than those of the other Levantine waters. The stream of Nab'l Berdauni measures an average of 8°C, with a winter minimum of 5°C (Alouf, 1982); in the streams feeding lake Yamoune, the temperatures fluctuate between 5 and 12°C; in the Upper Orontes the range of temperatures is between 10 and 15°C (Moubayed, 1986). In the rivers of the Lebanese mountains which are snow-covered for many days and have below-zero temperatures for more than three month a year, we find the main concentration of the Palearctic aquatic fauna of the Levant.

The two coastal rivers Damour and Aouali have very large, Mediterranean-type seasonal hydrographic fluctuations. Accordingly temperature variations are very large, i.e. from 8°C to 26 or even 29°C. Still, the very rapid flow (Damour has an average slope of 3.3 m/km and Aouali of 2.2 m/km) and the well-developed riverine forest vegetation maintain a highly diverse stream fauna (Dia, 1983).

Analyzing the Plecoptera of Lebanon, Berthelmy and Dia (1982) consider the area to be a kind of peninsula of the Anatolian highlands. The aquatic fauna is dominated by animals which have a Caucasian or southeast European affinity (Dia, 1983; Moubayed, 1986). It is furthermore of interest that Dia (op. cit.) finds much resemblance between the insect fauna of the Lebanon and that of the large Mediterranean islands.

Based on the Trichoptera (Dia, 1983; Dia and Botosaneanu, 1983; Botosaneanu, 1973, 1984; Moubayed and Botosaneanu, 1985), it has been proposed to divide Lebanon into 3 biogeographical sub-provinces: 1. the upper courses of the Litani and of the Orontes; 2. the coastal rivers; and 3. the upper course of the Jordan. Moubayed (1986) considers that the main difference between the coastal rivers and the rivers of the Bekaa lies in the fact that the first have a more or less intact riverine forest, while the others have barren banks. Besides, the coastal rivers are not influenced by winter frost and the general climate is mild and more humid. Therefore, Plecoptera, Ephemeroptera and Trichoptera are more diverse in the short coastal rivers; for the Chironomidae and Oligochaeta, the open landscape of the Bekaa rivers does not pose a limitation. The only representative of the extremely cold-stenothermic Palearctic Blepharoceridae (Diptera) reported till now from the Levant, namely *Blepharicera fasciata* lives in the headwater torrents of the Damour and the Aouali (Dia, 1983).

Alouf (1982) associates the presence of the amphipod genus *Niphargus* with the karstic springs above 800 m altitude. The same author (Alouf, 1983) presents a very interesting analysis of the stream of Qab Elias, a tributary of the upper Litani.

Like many streams in this area, the spring and the upper reaches (between 1,250 and 1,020 m altitude) are temporary; the stream turns perennial only below this altitude. Interestingly, the temporary section is preferentially inhabited by Plecoptera, whereas in the perennial stretch, they are replaced by Ephemeroptera. Alouf considers that the capacity of the Plecoptera to survive in the hyporheic waters and their competitive disadvantage versus the Ephemeroptera which are unable to survive in temporary streams, explains this distributional pattern.

Dia (1983) classifies the insect fauna of the coastal rivers (especially the Ephemeroptera, Plecoptera, Elmidae and Trichoptera) into species which are typical for the upper, median and lower courses of the rivers.

Botosaneanu, the co-author of a well-known stream and river typology (Illies and Botosaneanu, 1963), tries to adapt this classification to the running waters of the Lebanon. In doing this, he is aware of the fact that the stream basins are strongly influenced, fragmented and often disjunctly connected because of the karstic nature of the river beds (Moubayed and Botosaneanu, 1985). In comparing the Lebanese streams with their European counterparts the stability of the water temperature and the well-oxygenated turbulent flow are deemed more important than the absolute temperature of the waters.

With these reservations, Moubayed (1986) presents a classification of the streams of the Bekaa: according to him, the Rhithral would extend between 1,150 and 900 m of altitude in the Litani basin and between 1,360 and 1,000 m in the Orontes basin. The Potamal of the Litani would thus start at about 900 m altitude. Moubayed (op. cit.) gives a first listing of the animals typical for the three different stream zones, i.e. the Crenal (spring zone), the Rhithral (the stream or brook zone) and the Potamal (the river zone). Among the Tricladida, *Crenobia* sp. is typical for the upper Rhithral, being followed by *Dugesia* and *Dendrocoelum*. The chironomids of the Diamesinae subfamily are confined almost exclusively to the Crenal and upper Rhithral. The author gives also a distribution of the Ephemeroptera, Trichoptera, Simulidae and Coleoptera, according to the stream zones. The species which are confined to the Crenal and to the Rhithral are as a rule Palearctic species. For instance, the Diamesinae, reach their southernmost limit in the Lebanese mountains. The Plecoptera are represented in Lebanon by 13–14 species (Berthelmy and Dia, 1982), whereas in northernmost Israel and on the Golan Heights, there are only 5 species (Bromley, in press).The streams of the Golan Heights (some of them temporary!) may have a winter temperature of 5°C. No Plecoptera have been reported from further south, until the mountains of Ethiopia.

The fish fauna of the Lebanon is very poor, composed of mountain cyprinids and cobitids. Two species of the genus *Noemacheilus* are typical for the strong-flowing Rhithral and Potamal, while another cobitid, *Cobitis* sp., lives in the slower-flowing parts of the Litani. *Phoxinellus* is represented by one species each in the Orontes and the Litani, namely *P. kervillei* and *P. syriacus*. Lake Yammoune is inhabited by the highly endangered endemic species *P. libani*. The typical

"Mesopotamian" element (see above, p. 126) is represented only by the extremely widespread *Capoeta damascina* and the endemic *Hemigrammocapoeta festai* (Moubayed, 1986; Krupp, 1987). The rest of the fishes of Lebanon are either euryhaline coastal fishes or introduced species.

The Mollusca of Lebanon are little known. It appears from the available data (Kinzelbach and Roth, 1984; Kinzelbach, 1987) that the Bekaa represents a discontinuity in the distribution of the typical Rift Valley forms. The streams of the Bekaa are inhabited by the most widespread aquatic mollusk types of the Levant, such as the bivalve *Corbicula fluminalis* and the snails *Theodoxus jordani* and *Melanopsis praemorsa buccinoidea*. Schütt (1983) adds to this list the snails *Valvata saulcyi* and *Bithynia phialensis*, both widespread Levantine species. Interesting is the isolated report of the western Palaearctic *Ancylus fluviatilis* from Lebanon and from the coastal stream of Nahr el Kelb (Syria). Schütt (op. cit.) also mentions several species of *Pisidium*, e.g. *P. annandalei, P. casertanum, P.*

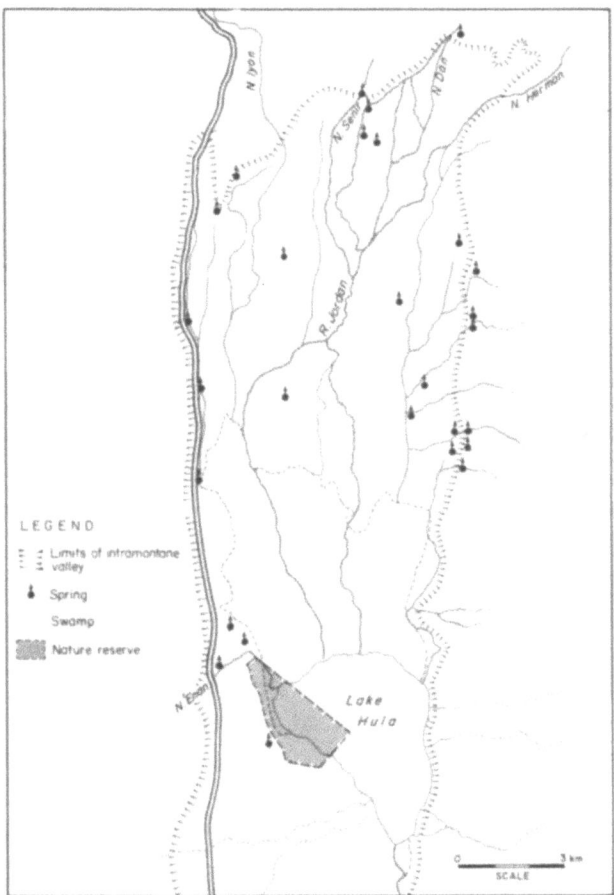

Fig. 6.13. Lake Hula before drainage works in 1951, its tributary streams and springs. The present nature reserve is also indicated (modified after Karmon, 1960; Paz, 1976 and data from the Hydrological Service of Israel).

personatum and *P. subtruncatum*, from a small stream from Mt. Hermon (Antilebanon). The last two species are considered by Schütt to be Palaearctic species. *P. personatum* is characterized as a cold-stenothermic crenal species, known also from the profundal of European lakes.

6.13. Lake Hula and the headwaters of the Jordan
by Ch. Dimentman and F.D. Por

The northernmost section of the Jordan Rift Valley is formed by an intramontane valley, the valley of Hula (Fig. 6.13). This valley contained a lake, known under various names such as Lake Semachonitis, Waters of Merom, Lake Huleh or Hula.

Lake Hula was surrounded by extended swamps and a number of strong karstic springs. The strongest spring of the valley, the Dan, forms the equally named river (Nahal Dan) which after a few kilometers converges with two other rivers which originate outside and above the Hula valley, namely Nahal Senir (Nahr al Hasbani) and Nahal Hermon (Nahr Banias). Their confluence forms the river Jordan. In recent times, after flowing some 11 km through the swampy lowland, the Upper Jordan discharged into and crossed Lake Hula.

The Valley of the Hula (see Karmon, 1960) is an area of approximately 175 square kilometers, situated at an average elevation of 90 m. It is bordered to the east and the west by mountain ridges which raise 500–1000 m above the valley floor; to the north and to the south the valley is confined behind ridges of Pleistocenic basaltic outflows, 130–260 m above the valley floor.

As shown by Picard (1963) and by Horowitz (1973, 1978) Lake Hula was formed in the early Pleistocene and underwent only small-scale "quantitative" changes since that time. It never turned into a saline lake and probably communicated for most of the time with the lower Rift Valley through a Proto-Jordan. Nonetheless, it exhibited considerable level fluctuations induced by changes in pluviosity, tectonics and associated basaltic flows (see above, p. 34). Accordingly, lacustrine sediments alternate with peat, i.e. open lake, with swamp situations. According to Picard (op. cit.) the late Pleistocene is represented by the sequence of the "Lower Lacustrine Beds", the "Main Peat Unit", the "Upper Lacustrine Beds" and the recent "Surface Peat Unit". They represented according to Ehrlich (1973) a sequence of alkaline and acidic conditions. The Upper Lacustrine phase represented a deep lake which probably reached +200 m level and terminated around 18,000 B.P., together with the end of Lisan Lake. This so-called "Gadot Lake" eventually covered most of the Hula valley and turned the spring of Dan, situated at +200 m, into a near-shore spring. The receding lake left behind an area of spring and river swamps which deposited extendend areas of spring travertine. Subsequently different streams, N. Iyon, N. Senir, N. Dan and N. Hermon united to form the Jordan (see Por et al., 1986). At the outlet from the Hula, the second stretch of the Jordan also had an eventful history. The falling of the base level of Lake

Kinneret to −200 m probably very much increased the capacity of the Jordan to erode its bed through the basaltic barrier, ultimately draining much of the old lake Hula.

Even historically recorded events had a considerable impact on Lake Hula. Karmon (1960) considers that the building around 1260 A.D. of the bridge at Benot Yaaqov on the Jordan outflow by the mameluk sultan Baibars led to a considerable narrowing of the river. As a consequence, the lake and especially the surrounding swamps extended again almost to the northern edge of the valley. As a result human settlement in the malaria-infested valley ceased almost completely. Only after 1830 and 1840, and especially towards the end of the nineteenth century, some artificial widening of the Jordan outflow brought about the situation recorded by the first scientific expeditions.

During the first half of the present century, Lake Hula occupied an area of 13–14 square km and the surrounding swamps an additional 37–49 sq. km. This represented more than one third of the total valley bottom.

Shortly after World War II an ambitious project of drainage was put into action. By 1951 most of the water of the Jordan headwaters was diverted from the lake into drainage canals. By 1958 the totality of lake Hula was drained. From the beginning a small Nature Reserve was set aside (Fig. 6.13), which gradually came to depend more and more on drainage from fish ponds (see Karmon, 1960; Dimentman et al., in press). The result was that many of the original biota died out in the reserve: though it fulfilled its original role as an important stopover for migrating birds, the aquatic fauna of the Reserve suffered very much. Efforts have been made during the recent years in order to improve somewhat the water quality of the Reserve.

As a consequence of the drainage, several changes occurred in the Hula valley and in the regime of the Middle Jordan. As reviewed by Inbar (1982), there was no decline in the amount of water input to Lake Kinneret: loss by modern agricultural and urban use about equals the evaporative loss of the old Lake and swamps. However, the winter floods which used to be buffered by the Lake now became a dominant feature of the Middle Jordan: in consequence the braided structure of its outflow from the drained Hula valley frequently changes. The new hydrological regime also led to an increased deposition at the outflow in Lake Kinneret, and a 750 m broad delta has been rapidly built.

The remaining freely flowing stretches of the rivers Dan and Hermon were recently exposed to further pumping projects. A Nature Reserve was established on the upper course of the Dan in 1969 and soon after this on the Hermon. Yet, facing further water demands, a survey on the fauna of the Dan was carried out (Por et al., 1986). For N. Hermon only scattered faunistic data are available. The situation of Nahr al Hasbani (Senir) is even worse, the river being strongly polluted already in its Lebanese stretch.

Recently, a painstaking effort has been undertaken to make a bibliographic reconstruction of the fauna of Lake Hula (Dimentman et al., in press). Though several expeditions went to the Lake before its drainage, the faunistic data were never assembled and many collections have not been worked out. In addition, data from the present Nature Reserve were also utilized; this was extremely important for taxa in which no pre-drainage collections were made.

From these monographic efforts the knowledge has been gained that the headwaters of the Jordan, Lake Hula, its surrounding springs and swamps, functioned as a system of interlocked aquatic environments, inhabited by an extremely diversified aquatic fauna. Four main features characterized this Eden of the Levantine aquatic life: 1. the already mentioned totally fresh waters; 2. the wide range of absolute temperatures and the graduations from eurythermy to stenothermy; 3. the variety of streaming waters, open lake and swampy environments; and 4. the rich aquatic vegetation of the Lake and swamps as well as the undisturbed riparian vegetation of the streams. As a result, the aquatic fauna of the Hula Valley presents the highest reported diversity of aquatic fauna in the Levant. More than 150 animal taxa identified mostly to the species level, are on record from the torrential part of River Dan alone (Por et al., 1986). An extremely rich aquatic vegetation has been similarly reported from this river: diatoms (Ehrlich, 1983), other algae (Eren and Dor, 1983), aquatic Bryophyta (Herrnstadt and Heyn, in press) and aquatic phanerogams (Lipkin, 1983). From the Hula, more than 500 species of aquatic invertebrates and fishes are listed (Dimentman et al., in press). There is very good information about the aquatic plants of the Hula, too, both before and after the drainage (Ehrlich, 1973; Jones, 1940; Zohary and Orshansky, 1947; Shmida, 1975).

Nahal Dan, a river formed by an extremely strong karstic resurgence is characterized by a 5 km long stretch of almost completely sternothermic water with a temperature of 15.5°C. This is due to the abundant flow of 250 million cubic meters/year and the fact that there is only a restricted seasonal fluctuation of the discharge values. The Dan has an extremely turbulent upper flow, often reaching a current velocity of 200 cm/s. Oxygen saturation is constantly around 95%. The fauna of the Dan is therefore a typical stenothermic rhithral fauna, the overwhelming majority of the species being Palaearctic species.

The diversity of these mostly rhithral and Palaearctic taxa needs to be mentioned: 4 species of Tricladida ; 5 species of Oligochaeta Naididae; 3 species of Plecoptera; 11 species of Ephemeroptera; 16 species of Trichoptera; 8 species of Coleoptera Elmidae; 3 species of the hemipteran genus *Gerris* and 2 species of *Sigara*; 6 species of Copepoda Harpacticoida of the family Canthocamptidae; 6 species of the bivalve *Pisidium*; 3 species of loaches (Cobitidae).

About 20% of the species living in Nahal Dan are known in Israel till now only from the three sources of the Jordan. Almost half of the species are limited to the Upper Galilee and to the Golan Heights (i.e. north of the " Nehring Line ", see below, p. 153). No doubt these populations are the southernmost representatives of a Lebanese mountain fauna. A further case to this point is the report of the subterranean amphipod *Niphargus nadarini* in the spring of Dan and one near-shore spring of Lake Hula; this is probably a Lebanese species and the localities are the southernmost reports of this cold-stenotherm genus.

Only isolated species of Ethiopian insects are on record like the hemipterans *Belostoma cordofanum* and *Micronecta scutellaris*, the dragonfly *Pseudagrion syriacum*, the chironomid *Polypedilum aegyptium* and eventually the ephemeropteran *Afronurus kugleri*).

Most important of all, with the exception of the flatworm *Dugesia biblica* there are no representatives of the euryhaline fauna which is so characteristic for the Jordan system.

The river-resurgence of N. Hermon, has a less stable seasonal flow. Besides, the spring is found at an altitude of +400 m., i.e. the stream was a permanent feature even when the Dan became a near-shore or subaquatic spring of the "pluvial" Lake Hula. N. Hermon has therefore a few euryhaline, or Jordan, species, which are missing from the Dan, such as the harpacticoid copepod *Nitocra incerta* and the cichlid fish *Tilapia zillii*. In general, however, the two Jordan headwaters seem to have a very similar fauna.

Lake Hula can be characterized as a shallow lake with extreme seasonal thermal fluctuations: from 6–7° to 36°C. The many littoral springs which were in direct contact with the lake or with the swamps presented fairly constant temperatures of 19° to 21°C. In the past when Dan was a near-shore spring, even colder stenothermic environments were provided. As a result, Lake Hula was inhabited primarily by cold-loving (or cold-resistant) basically Palaearctic species: during the high summer temperatures, the littoral springs provided cool refugia (Dimentman et al., in press).

However, Lake Hula also contained an important euryhaline and an Ethiopian element. For instance among the 19 species of fish, there are 4 tilapiine cichlids and also *Garra rufa* and *Aphanius mento*, lived together with 5 species of Palaearctic Cobitidae. The thermophilic species used the stenothermic springs in their turn, this time as refugia in the winter.

Among the smaller crustaceans there are several species of Ethiopian affinities, known also from Lake Kinneret, such as *Bosmina longirostris*, *Ceriodaphnia reticulata* and *Afrocyclops gibsoni*. To this we can add also *Tropocyclops confinis*. *Bosmina longirostris* is also known from Pleistocenic sediments of the lake (Ohlhorst et al., 1982). Among the aquatic insects there are many southern species, especially among the more than 20 species of Culicidae. Linnavuori (in litt.) identified from Lake Hula such southern species of Hemiptera as the Egyptian belosomatid *Limnogeton fieberi*.

The rich aquatic vegetation of the Lake was and still is to some extent, extremely hospitable to aquatic insects. Dimentman et al. (in press) list over 90 species of aquatic Coleoptera, based also on post-drainage material.

As described by Wewalka (1986) and by Jäch (in.lett), some of these species are new to science and possibly endemic.

The problem of the Hula endemics deserves a more detailed discussion in order to better evaluate the destructive impact of the lake drainage. No doubt an immense number of species did not survive in the extremely small Reserve. The Reserve

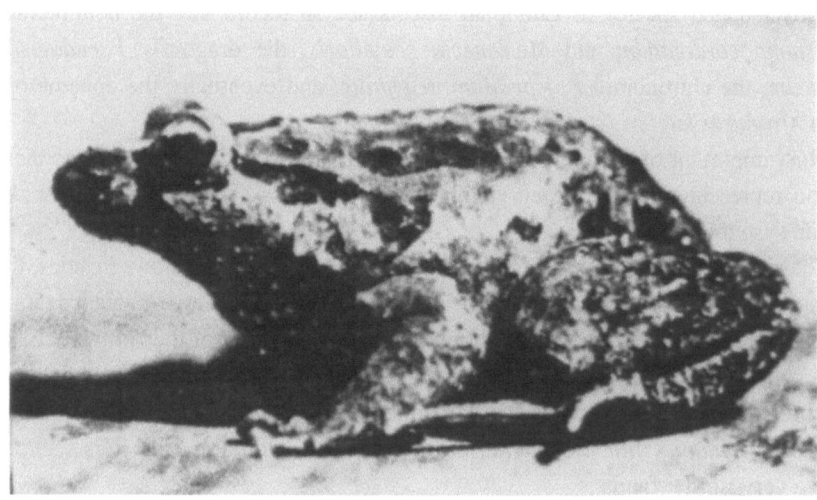

Fig. 6.14. Discoglossus nigriventer – Endemic frog of the pre-drainage Lake Hula (courtesy H. Mendelssohn).

does not have its natural water sources and is fed by nutrient-loaded water from the surrounding fish ponds; moreover, it had lost contact with the surrounding stenothermic spring refugia and the submerged and floating vegetation has been drastically reduced. However, many species which disappeared from the lake still survive in the now isolated springs, in the swampy area of the Bet Zaida valley on the northeastern shores of Lake Kinneret and indeed in the Lebanon. To what extent this is correct for all the old inhabitants of Lake Hula is still subject for further research.

However, in some cases, where knowledge is sufficient, the situation is already clear. The frog *Discoglossus nigriventer* (Fig. 6.14) has not been reported since 1955 (Steinitz, 1955). On the other hand, the presumed endemic subspecies of fish *Tristramella simonis intermedia* (Steinitz and Ben-Tuvia, 1960) has recently been found in the Bet Zaida swamps (Ben-Tuvia, pers. comm.). The damselfly *Pseudagrion torridum hulae* is eventually a relic subspecies of the Nilotic nominate species (Dumont, 1973, 1975, in press; Schneider, 1987); this subspecies survived in the Hula after drainage. The only dragonflies eventually disappeared with the Hula are: *Rhyothemis semihyalina syriaca* and *Urothemis edwardsi hulae*.

To conclude, the watery environments of the Hula valley, i.e. the lake and the surrounding running waters provided the last southern outpost of a normally diversified stenohaline and stenothermic Palaearctic freshwater fauna. Though this fauna can still be found in other more northern areas of the Levant or even outside the province, the diversity which was present in the Hula basin was unique to our area. As shown by Dimentman et al. (in press), under certain conditions of correct water management, many of the original species could recolonize the impoverished Nature Reserve of the Hula in the future.

6.14. The limits of the Palearctic advance

South of the area of the valley of Hula, the Palearctic influence is severely restricted. The area where this radical impoverishment occurs cuts across the southern Levant corresponding more or less to the "Nehring line" proposed by Por (1975b) (Fig. 6.15) for the Palearctic terrestrial biota. This "aquatic Nehring line" (see also Por, 1984b) is situated south of the Golan streams which flow into Lake Kinneret, south of Lake Hula and of the small coastal streams of the western Galilee.

A transitional area which occupies much of the territory of Israel is an area of recent advance or relic survival of Palearctic aquatic species, a situation which probably best expresses the fluctuating climatic history of the southern Levant.

Few of the most resistant Palearctic species can still be found in the springs and the short streams of the Cis- and Transjordanian mountain ranges. These streams have short and seasonally unpredictable perennial flow; they often retreat underground and even their springs can stop flowing for some time. Temperatures measured in the springs are around 21°C and hydrographic parameters in the short and/or semipermanent surface flow, are widely fluctuating.

Nevertheless, the number and the density of these rudimentary stream basins is so large that the survival of an impoverished Palearctic aquatic stock was possible. In Israel, south of the Nehring Line there are 16 streams which flow to the Mediterranean and 7 streams which flow to the Jordan system. The Transjordanian tributaries are in general more abundantly flowing than their Cisjordanian counterparts.

Few drainages carry a permanent stream from the source to the outflow into the Mediterranean, the Jordan or the Dead Sea. They tend to have short mountain stretches and then have a seasonally dry middle-stretch and permanent flow is found again near the sea shore. These coastal stretches are often brackish estuaries (see p. 105). Recent use of these stream beds as sewage carriers gives today a deceptive image of their original hydrography.

The Cis-Jordanian streams are inhabited only by one species of loaches (Cobitidae), namely *Nun jordanicus.* (Banarescu et al., 1982). This species is merely found in the streams which reach the Jordan system, the southernmost being Wadi Kelt. On the Trans-Jordan side, even rivers reaching the Dead Sea have their fish fauna. *Nun jordanicus* is found only in the headwaters of Nahal Yarqon a westward flowing coastal river. With these exceptions in mind one can say that the area of the residual Palearctic fauna has no freshwater fishes.

As for the invertebrates the helping hand of the water supply systems of the past millenia might have contributed in order to vouchsafe survival and recolonisation. In some cases, species considered to be "subtroglophilic" by Botosaneanu (1974), such as certain Trichoptera (species of *Micropterna* and of *Mesophylax*) can survive the hot months when there is no free-flowing surface water, as winged

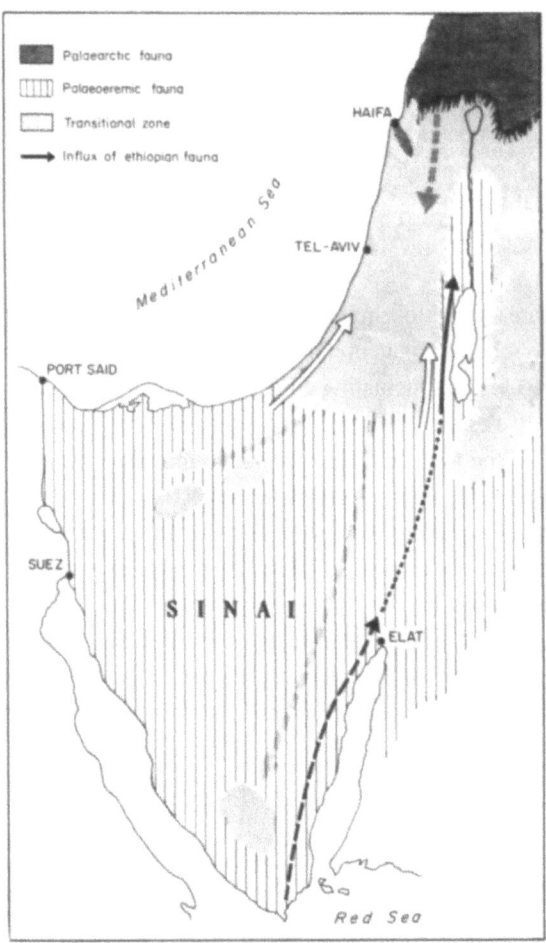

Fig. 6.15. Limits of distribution of terrestrial Palearctic biota in the southern Levant. The mainly aquatic Ethiopian influence along the Mediterranean shore has not been put in evidence (from Por, 1975b).

dormant imagines hidden in the humid spring caves. Many other species of this resistant stock inhabit the natural or excavated spring caves: In this dark and cool environment, there is a permanent trickle of water, and restricted aquatic or wet soil environments remain till the next winter.

The group of hardy Palearctics already mentioned elsewhere (Por, 1975b) contains among others the following species: *Atrioplanaria aquaebellae, Dugesia biblica* (Bromley, 1974), *Pisidium casertanum, Melanopsis praemorsa buccinoidea, Dina shtschegolewi, Proasellus coxalis, Gammarus syriacus, Eucyclops serrulatus, Canthocamptus microstaphylinus, Bryocamptus minutus* and *Attheyella crassa*. The last three species are canthocamptid harpacticoids (Copepoda). North of the Nehring line, there are several more species of this family, among them the genera *Elaphoidella* and *Moraria* (Chappuis, 1955; Por, 1983). The insect fauna of the impoverished mountain springs of Israel includes as a rule only two species of Ephemeroptera, namely *Caenis macrura* and *Cloeon dipterum*.

The fauna of Trichoptera is also impoverished, as well as that of several other orders of aquatic insects. Plecoptera are not found south of the Nehring line (Fig. 6.15).

The residual Palearctic spring and stream fauna reaches the southern limit of its present expansion along a transverse line called for the terrestrial fauna the "Bodenheimer Line" (Por, 1975b). Indeed there is no complete superposition between the biogeographic discontinuity of the terrestrial and aquatic faunal limits. The area, however, is grossly identical: the foothills of the Judean mountains in Israel and eventually the area between Shubaq and Petra in Transjordania. The limits of the Palearctic advance along the Trans-Rift mountains are not yet clear. The possibility of occasional expansion events along the mountain chains which accompany the Red Sea has to be taken into account. Tchernov (1988) mentions the case of the cobitid fish *Noemacheilus abyssinicus* as an isolated case of a Paleartcic fish occurring in the Ethiopian Lake Tana.

A few isolated areas of spring-oases situated south of the Bodenheimer line contain still several Palearctic species. Two examples will be brought. First, the group of abundant springs at En Qudeirat, in north Sinai, about 80 km south of El Arish. There, the diversity of Palearctic species is still relatively high: as an example, this is a southern outpost of the flatworm *Dugesia biblica*, of the freshwater crab *Potamon potamios* and of the isopod *Proasellus coxalis*. There is a distinct possibility that during more propitious climatic phases these oases were included in the area of the continuous Plaearctic distribution.

The second example is that of the group of springs in the mountains of southern Sinai, which feed several stream-oases, such as Wadi Feiran, Wadi Shaag, Wadi Taal, Wadi Arbain, etc. Several of these springs and spring brooks are at high altitudes of over 1500 m (the highest peak of S. Sinai, Gebel Katarina is 2642 m heigh) and water temperatures are around 16°C.

An isolated Palaearctic fauna can be found there, such as the canthocamptids *Attheyella crassa* and *A. naphtalica* (Por, 1983c); the above-mentioned trichopteran *Mesophylax aspersus* (Botosaneanu, 1974); the culicids *Culex mimeticus* and *Uranotaenia unguiculata* (Margalit and Tahori, 1973); the hemipterans *Sigara lateralis* (Linnavuori, 1964) and *S. marginata* (Brown, 1951). The two ephemeropterans widespread in Israel, namely *Cloeon dipterum* and *Caenis macrura* are also reported. There are also two leeches, namely *Helobdella stagnalis* and *Dina sp.* which are missing in the intervening areas of the Negev (Bromley, in press).

The majority of the insect fauna, however, is of the Ethiopian and Eremic type. This is especially the case for the 23 species of Odonata from Sinai, reported by Dumont (in press).

It appears that the mountain area of Sinai with its springs behaves biogeographically as an island, where only aerial distributing or passively carried species can reach. The understanding of the aquatic molluscan fauna of Sinai poses difficulties, however: it contains besides the euryhaline species *Pseudamnicola solitaria*, also the freshwater species *Melanoides tuberculata*, *Physa subopaca* and *Valvata saulcyi* (Tchernov, 1971b). Very old distributional patterns or passive transport are the two possible explanations.

The large spring of En Avedat situated about 60 km southwest of the Dead Sea, is an important oasis in the basin of Nahal Zin, once a tributary stream of Lake Lisan. Although isolated presently from the Jordan-Dead Sea system, the perennial En Avedat and its short stream flow, are inhabited by euryhaline species of the Jordan Rift Valley.

6.15. The ephemerous waters of Israel and the Levant
by Ch. Dimentman and F.D. Por

Rain pools and brackish or saline temporary waters are common in the Levant. The earliest data on their fauna are those by Baird (1859) and by Barrois (1892). From Jordan, mention should be made of the more recent contribution by Löffler (1967).

More complete and intensive studies on the rain pools and their fauna have been carried out in Israel (Yaron, 1964; Hartland-Rowe, 1967; Dimentman, 1976, 1981; Dimentman and Por, 1982).

Dimentman analyzed the fauna of anostracan crustacea from 146 rainpools in Israel, the Golan Heights, the areas of Judea and Samaria as well as the Gaza strip (Fig. 6.16). The five species identified, namely *Branchinecta ferox*, *Branchipus schaefferi*, *Streptocephalus torvicornis*, *Chirocephalus bairdi* and *Ch. neumanni*, probably represent the totality of the fresh-water anostracan fauna of Israel and the studied adjacent territories. Besides the anostracans, three taxa of Notostraca are on record, namely *Lepidurus apus apus*, *L. apus lubbocki* and *Triops cancriformis* (Fig. 6.17). The conchostracan fauna of our area, preliminarily identified according to old literature, is badly in need of a revision. Three taxa are known: *Cyzicus* sp., *Eocyzicus* sp. and *Lynceus* cf. *brachyurus*.

The fauna of the diaptomid copepods in the rainpools of Israel, Jordan, the Golan Heights and Sinai is restricted to two subspecies of *Arctodiaptomus similis*, *Hemidiaptomus gurneyi canaanita* and *Neolovenula alluaudi* (Dimentman and Por, 1985).

Among the cyclopoid copepods *Microcyclops minutus* is the predominant species, though the whole group of genera which is contained under the large generic name *Microcyclops* is in need of revision.

The Cladocera are represented in the rainpools by a variety of species of the genera *Moina*, *Daphnia*, *Macrothrix* and by genera of the Chydoridae. One of the

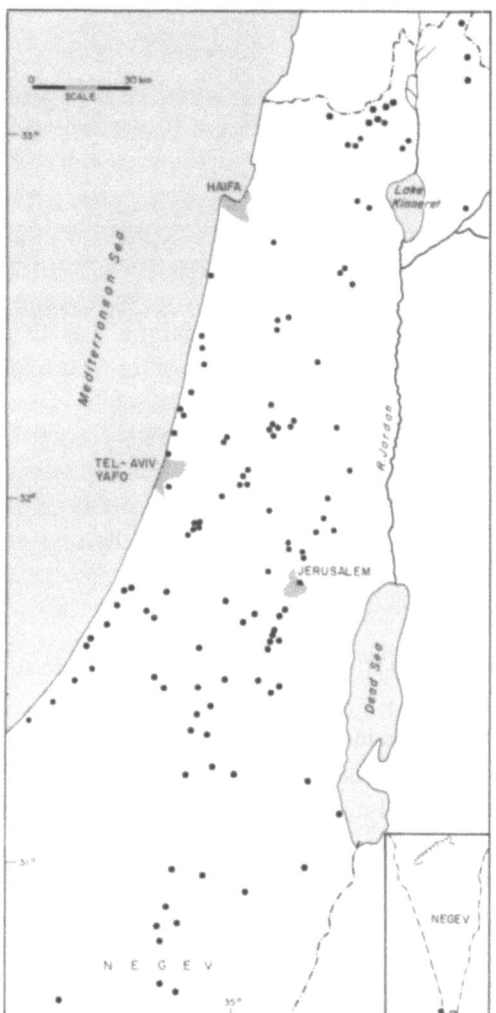

Fig. 6.16. Temporary waters in Israel and surrounding areas, inhabited by fairy shrimps (Anostraca) (after Dimentman, 1976).

local rainpool organisms in Israel is *Daphnia atkinsoni*. Unfortunately, the rainpool Ostracoda and Rhabdocoela (Platyhelminthes) have not been studied yet.

From the map attached, (Fig. 6.16) it appears that the Anostraca are distributed in accordance with the hypsometric curves and also with a complex of actualistic environmental factors, such as length of existence of the pool, depth of the pool, nature of the substrate, microclimate, adaptations of the resistance "eggs", and inter-specific relations.

One of the most intriguing and still not well understood feature is the exclusive and geographically limited presence of *Neolovenula alluaudi* in the rock pools of the Negev and northern Sinai, tens of kilometers south of the limits of the typical crustacean rainpool fauna of Israel. The two subspecies of *Lepidurus apus* are

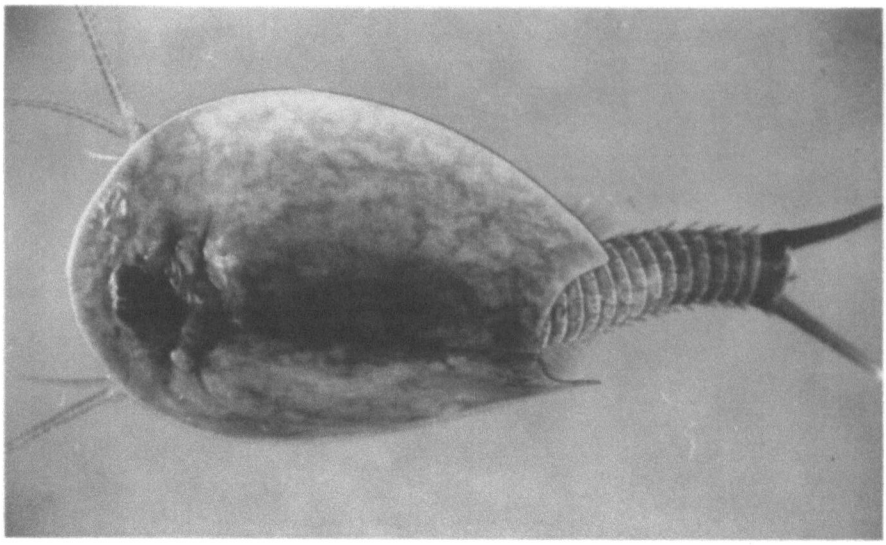

Fig. 6.17. The tadpole shrimp *Triops cancriformis* (Notostraca) (photo D. Darom).

vicariant: ssp. *lubbocki* lives in the more mild climate of Israel and Cis-Jordania, while ssp. *apus* lives in the more continental climate of the Golan Heights. In general, there exists a typical rainpool fauna of the arid areas, such as *Branchipus schaefferi* (Negev and northern Sinai), *Eocyzicus* sp. (northern Sinai), *Triops* sp. (northern Sinai and east Jordan), the above-mentioned *Neolovenula* and another diaptomid *Metadiaptomus chevreuxi* (Trans-Jordan) (Dumont, 1979).

Genus *Chirocephalus* is represented in the Levant by several endemic species, such as *C. bairdi* and *C. neumanni* in Israel and *C. appendicularis* in Lebanon.

Though we have little informative data on the Diaptomidae of Lebanon and Syria, it is important to note that the Israeli fauna appears as extremely impoverished, when compared to the very diversified fauna of Anatolia. Future investigations will have to reveal where in the Levant the impoverishment in the diversity of these typical rainpool crustaceans occurs.

It is interesting to mention that there exists a non-specific rainpool fauna, like the harpacticoid *Canthocamptus microstaphylinus* which has resistance eggs and can survive drought. However, this species is found in temporary pools which are in contact with springs or discrete groundwater sources. Likewise, the bryozoan *Plumatella repens* is present in temporary pools which, either because they are deep enough or because of the fact that they are fed by trickles of groundwater maintain a certain soil humidity in which the statoblasts can survive. A typical example of such "mixed" temporary rainpools is Bab el Hawa of the Golan Heights. *Arctodiaptomus similis*, too, can be found in other types of standing waters, sometimes appearing even in Lake Kinneret.

The amphibian fauna of the temporary pools of Israel contains *Triturus vittatus*,

Salamandra salamandra, Bufo viridis, Hyla arborea, Pelobates syriacus and rarely *Rana ridibunda*. As shown by Yom Tov and Mendelssohn (1988), many rainpools in Israel have ceased to exist because of urban, agricultural and road development. The populations of *Triturus, Salamandra and Pelobates* are especially endangered. One might add that the conchostracan *Lyncaeus*, mainly found in the Coastal Plain, is also severly endangered. Several small Nature Reserves have been established in order to maintain the triton and salamander populations.

A Nature Reserve of extreme importance is the ancient rainwater cistern Birket Mamilla, today in the center of downtown Jerusalem, which is the "locus typicus" of 4 valid species of crustaceans described by Baird (1859)! With the drainage of the standing waters of the coastal plain of Israel, several species of large aquatic coleopterans formerly reported from there, have also disappeared (Jach and Ortal, 1988). These are species of Ethiopian affinities which have not been found yet elsewhere in the Levant.

An anostracan which is typical to saline desert pools is *Branchinella spinosa*, which is on record from Qa el Azraq in Jordan (Scates, 1968) and from other desert pools in Syria.The hypersaline waters of the Levant are inhabited by several species and strains of parthenogenetic and amphimyctic *Artemia* (Dimentman, 1980). None of the species of hypersaline Diaptomidae have been found in Israel so far: they are present, however, in Anatolia.

6.16. Polluted and manmade waterbodies
by Ch.Dimentman and F.D.Por

As mentioned above, the human impact on the continental waters of the Levant is far-reaching. This has been amply discussed above, especially with respect to the hydraulic changes. Pollution has modified the hydrology and biota of the flowing waters as many coastal streams turned into more or less purified sewers. A typical seasonal stream of the Judean Mountains, Nahal Soreq, turned into a permanent sewage stream upon receiving the partly treated sewage of Jerusalem. Bromley and Por (1975) have discussed the biota of the self-purification succession in this sewage stream. A typical biotic sequence of algae and invertebrates was described and tentatively compared with the "saprobic" categories accepted for the European polluted waters. Along a 40 km flow of the sewage carrying wadi of Nahal Soreq, a degree of self-purification was reached, corresponding to the beta mesosaprobic level. The last collecting stations were characterized rather typically by the isopod *Proasellus coxalis* and by the ostracods *Herpetocypris chevreuxi* and *Eucypris clavata*. Such a "Proasellus zone" or "Ostracode zone" characterizes also the recovery of other sewage stream biota in Israel.

The example of Nahal Soreq is widespread. Many coastal streams which exhibited only temporary flow for most of the year, turned into perennial sewage-water streams (Fig. 5.13). Recently, with increasing use of the sewage in treatment ponds, this trend starts to revert. Needless to say, however, that the damage has already been done.

Even the Lower Jordan, south of Lake Kinneret had its flow regime artificially changed. Already in the 1930's with the building of the Power Plant on the Jordan outflow and afterwards when the Israeli National Water Carrier and the Jordanian Eastern Ghor Channel became operational, the Jordan lost about 90 % of its total flow (Ortal and Por 1978; Por and Ortal, 1986). A series of dams and barrage lakes especially on the Jordanian side, modified very much the aquatic landscape of the area (Figs. 6.18 and 6.19).

Besides the drastic reduction of flow, the Jordan which usually presented a winter flood regime was turned into a strange stream which has an artificial summer flow.

This is precisely the time of the year when irrigational water is needed for the adjoining fields (Ortal and Por, 1978). Only in extremely rainy winters in which the run-off exceeds the storage capacity of Lake Kinneret does the Jordan revert for a short time to its original flood regime. In such cases the effects on the Dead Sea level are immediately felt.

For similar reasons of water storage a number of small dams were built on the

Fig. 6.18. King Talal Dam on Nahr Zarqa, Jordan (from Khouri, 1981).

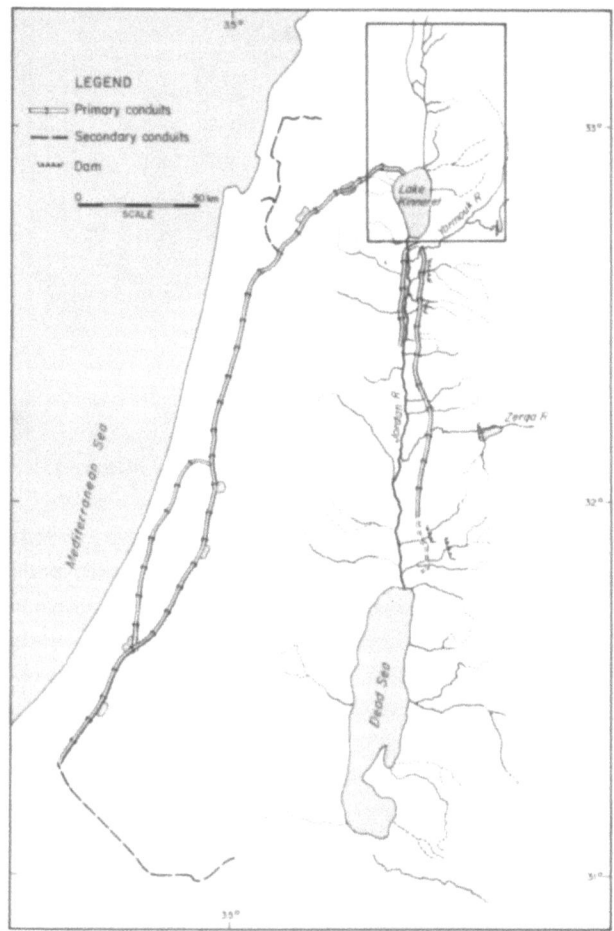

Fig. 6.19. Major hydraulic changes related with River Jordan (original, based on Leventer, 1984 and Atlas of Israel, 1985). See Fig. 6.20.

fast flowing streams which descend from the Golan Heights to Lake Kinneret (Por and Ortal, 1986) (Figs.6.19 and 6.20).

During the two last decades the number of water reservoirs used for irrigation has increased considerably in Israel. Yom Tov and Mendelssohn (1988) mention that by 1984 there were already about 260 such artificial bodies of standing water. This is, of course, in addition to the many hundreds of fish ponds and reservoirs of drinking water. Most of these reservoirs are typically small, not exceeding 0.10 square kilometers.

The source of the water and the use of it are different. They are filled with rainwater drainage, spring water, phreatic water from drillings, water from Lake Kinneret brought by the National Water Carrier and finally sewage water. In general, the water accumulated during the rainy season is used in the summer or early autumn for the irrigation of industrial crops.

Fig. 6.20. Recently built dams on the streams of the Golan Heights (original, based on map by Tahal Ltd. Haifa). (Insert on map 6.19.)

Klein (1988) indicates that in Israel about 100 million cubic meters per annum of flood water are stored in some 160 reservoirs. Moreover, 90 million cubic meters per annum of sewage water (out of a total of 240 million cubic meters) are used for irrigation, after secondary treatment in reservoirs.

For the aquatic fauna of Israel, an entirely new and widespread type of environment has been created, which is represented moreover in all the climatic regions of the country. One of these small-scale biogeographic modifications produced is the fact that several species, typical for Lake Kinneret, spread over the country, using the water pipes. Considering that Lake Kinneret water is distributed as far as the northern Negev, the long-term effects might be considerable (Fig. 6.19).

Leventer (1984) has summed up the diversity of the species found in the large reservoirs of the National Water Carrier. In general, there is a large degree of similarity in their biota, irrespective of the geographic location. The macrophytic vegetation is dominated by *Chara vulgaris, Najas marina* and four species of *Potamogeton*. The typical dinoflagellate of Lake Kinneret *Peridinium* does not survive in the reservoirs. The crustacean zooplankton includes among others the cladoceran *Bosmina longirostris* and the African cyclopoid *Mesocyclops ogunnus* (see p. 126). Several gastropods are found in all the reservoirs: *Melanoides tuberculata, Melanopsis costata* and *Lymnaea auricularia*, while the bivalve *Corbicula fluminalis* did not advance far from Lake Kinneret. However, several species from the lake are living and presently reproducing in the reservoirs. Such are *Tilapia zillii, Acanthobrama terraesanctae* and *Clarias gariepinus*. According to Leventer (op. cit.) they reached all parts of the water-supply system. Most interesting is the presence even of *Atyaephyra desmarestii* in reservoirs distant from the Lake Kinneret intake.

Several species of fish alien to the local fauna were introduced in the system of reservoirs of the National Water Carrier, in order to act in biological control of algal and zooplankton blooms. Such are the cyprinids *Hypophthalmichthys molitrix, Aristichthys nobilis, Ctenopharyngodon idella* and *Myllopharyngodon piceus* (Leventer, 1984). The introduction of the peixe-rey *Basilichthys bonairensis* did not prove too successful.

Species of fish which were succcesfully introduced in culture are the carp *Cyprinus carpio* (in the 1930's) and the rainbow trout *Salmo gairdneri* (in the late 1940's).

The carp is now producing some 60% of the fishpond yields in Israel and the rainbow trout is cultivated in the cold-stenothermic River Dan. Although the tanks are stocked with trout fingerlings imported from abroad, there is a strong indication that free breeding occurs in the headwaters of the Jordan. In an important summing up on this subject, Ben-Tuvia (1981) reports several less welcome introductions of economic fish: the tilapia *Sarotherodon mossambicus* which might create unwanted hybrids with local cichlids; *S. niloticus*, an occasional visitor in the coastal streams, is being used to produce male-only hybrids with *S. aureus*; finally *Tilapia zillii* has been introduced by the Jordanian Government into the springs of En Faskha and to El Azraq. Several fish farms in Israel are cultivating the prawn *Macrobrachium rosenbergii*, another introduced species. The killyfish *Gambusia affinis* has been introduced to Israel in the 1920's, in order to fight the malaria mosquito.

Postface: The Legacy of Tethys – Or the Guise of a Conclusion

Every historical approach is inescapably attracted to identify nostalgic "golden ages". For the marine biogeographer, the Tethys sea in the shape of a world-encompassing tropical sea represented such a golden age. Starting with a gulf-like indentation in the late Paleozoic Pangea (the "Paleotethys"), it gradually separated the southern from the northern continents (the "Neotethys").

Geologists strike even more romantic notes when speaking of the wife of Okeanos. Sengor (1985) asked in the title of a study "How many wives did Okeanos have?" and McKenzie (1987) speaks of "Tethys and her progeny". A not less scholarly study by Carey (1987) is entitled "Tethys and her Forbears". In his paper, Carey postulates the existence of as many as three successive "Thethyses": a Riphean Precambrian one, an Early Paleozoic Caledonian one, and the most recent post-Permian one. For this author, Tethys is a repetitive phenomenon, "a pan-global equatorial geosyncline, with onset, life-span and climax within one era...". Do the Red Sea and the Suez Canal herald a future fourth Tethys?

Tethys in its Cretaceous climax is seen as the cradle of the most diversified aquatic fauna of all times. The demise of this golden age started with the movement of the African plate towards the Eurasian one. This movement caused the uplifting of the Levantine landbridge between the two continental plates. The resulting marginal sea, the Mediterranean, is the result of the different small-scale movements, lateral and frontal, between the two continents and their micro-plates. In this process the only area which possibly represents a persistent remnant of the Paleozoic Tethys is the Levantine Basin.

Nothing is known about the river systems which emptied into the Tethys in the Levantine area. It appears that the uplifting of the land bridge was coeval with the start of the Miocenic desertification trend. It turned the Mediterranean and its southern gulf, the fledgling Red Sea, into high-salinity, metahaline seas whereas the Levantine rivers and lakes faced ephemerous cycles of salination, drought and renewal. The large Mesopotamian riverine basin of the Tigris and Euphrates left behind by the receding Tethys became the natural center of origin for an aquatic fauna which advanced and retired along the ephemerous waterways.

At the end of the Miocene, during the several hundred thousand years of the Messinian phase, the Mediterranean turned into a huge endorheic system of terminal lakes. It was the ultimate demise of Tethys in our area.

The Pliocene marked the renewed activation of a transversal, south-north rifting movement. In probable connection with this, the Mediterranean, open again to the Atlantic, became a deep sea. The rifting was obviously responsible for the channeling of the Nile into the Levant basin. It created by longitudinal shear-movements the Levantine branch of the Rift Valley, its rivers, lakes and accompanying mountain ranges. In the south, the Red Sea deepened and opened up to the Indian Ocean. In the north, the Gulf of Aqaba came into being, whereas the older Gulf of Suez remained a shallow tectonic backwater.

In the wake of these tectonic revolutions, tropical biota entered again the Red Sea, within an isthmus-thin distance from the Mediterranean. Normal marine flora and fauna settled again in the Mediterranean, although a temperate one and a far cry from the old Tethyan tropical diversity. On land, high Cis-and Transrift mountains increased the pluviosity over the area and a series of strong artesian springs produced a steeple-chase of permanent Rift Valley rivers. The Levantine desert turned into the western branch of the "Fertile Crescent". Mesopotamian, African and Palearctic aquatic biota made forays into the newly-born riverine systems.

Upon this general trend of hydrobiological recovery of the Levant, the Pleistocenic glaciations superimposed their manifold fluctuating hydrologic and thermosaline regimes.

Further complicating the picture, at least in comparison with Europe and Africa, was the fact that the rifting tectonics continued during the glacial oscillations, exacerbating or mitigating the global climatic trends. With the last very active tectonism only some 10,000 years ago, technically speaking our area is still in geological turmoil.

The Mediterranean received influxes of Boreal fauna during the cold Glacial peaks and of Subtropical Senegalian fauna during the mild Interglacials. Some of the Boreal guests might have never reached the Levantine Basin, but the subtropical immigrants probably found a warmwater refuge in the Levant to survive till the next climatic optimum. Deeply disturbing the Levantine refugium where the repeated flood-like intrusions of freshwater, from the Black Sea or from the Nile: saline stratification and anoxic bottom conditions, resulted.

With little solace from river run-off, the Red Sea repeatedly experienced shrinking contacts with the Indian Ocean and turned into a metahaline waterbody; possibly only its deep northern Gulf of Aqaba, cooler and better supplied with freshwater run-off, maintained a reduced marine fauna.

Partial resettlement with tropical biota from the Indian Ocean followed in the wake of each deglaciation.

The continental waters of the Levant experienced full impact of the combination between glacial cycles and tectonic activity: Local mountain uplifting, extrusion of lava barriers or deepening of tectonic lake bottoms and even orogenic exposure of subterranean aquifers, combined with the global shifting of climatic zones. Only a very hardy, adaptable and easily dispersing freshwater fauna could survive when rivers unpredictably shrank and lakes turned into salty swamps. The last culmination of tectonic activity in the Levant occurred at a time when Modern Man was already present in the area and engaged in environment-modifying exercises. The Levantine man, like his Egyptian and Mesopotamian peers, dealt millenies ago in hydraulic engineering. On the small rivers of the Levant, the achievements were much less impressive; however, their impact on the triple mixture of biogeographic faunal elements was probably very strong. Artificial lakes, aquaduct systems and spring management had a serious impact on the intricate network of oases and desert rivers.

Many of the so-called endemics of the Levantine waters are probably relic occurrences and not "positive" evolutionary endemics.

In the Levant, long before Tethys the goddess was born, Man mastered Thiamat, the abysses of the sea. Ramses and Darius cut the Isthmus of Suez to allow navigation from sea to sea.

On the Mediterranean coast the Phoenicians built their artificial harbours and connected islands to the shore.

After the lull of the Middle Ages, water engineering started with renewed energy. The modern Suez Canal retraced for the first time a Tethys-like connection between the Mediterranean and the tropical world of the Red Sea. The successful reappearance of the tropical biota in the Levant Basin can be seen as a feat of biogeographical dimensions. After repeated encroachments, the flow of the Nile was finally obstructed by the Aswan High Dam and a further impoverishment in the biological productivity of the Levant Basin was eventually provoked.

Even much more destructive were the hydraulic projects on land. Venerable lakes, which successfully survived the Pleistocenic ordeals, were either annihilated, like Lake Amiq, or reduced to a shade, like Lake Hula. The Dead Sea shrank to nearly half its size and the obvious responsability of the hydraulic works on the Jordan which among others reduced Lake Kinneret to a mere reservoir, cannot be ruled out. But first and foremost, rivers were turned into sewage collectors and every drop of spring water is being used and reused.

Even though they carry used waters, many old temporary streams turned into permanent flows.

On the more positive side, fresh waters were channelled into areas where they were not available before and thousands of reservoirs and fish ponds were built. An entirely new waterscape appeared in the arid Levantine lands. The possibilities for biota ready to settle these new environments are plenty. Unfortunately, however, artificial introductions are often interfering with the process.

All in all, the legacy of Tethys is still alive. For the first time in perhaps 10 million years, the marine biota on both sides of the Levantine land bridge are meeting again.

To emphasize the uniqueness of this phenomenon is already commonplace. On land, the aquatic biota, though severely endangered, are being presented with new chances of expansion. Nature conservation, at long last, presents a few modest successes.

The Levantine province is not a area which can be considered as a "center of origin" for aquatic biota. It is rather an area in which dispersing populations of at least three different centers meet. It is not an ideal area for those who are looking for prevalent "vicariance" patterns: The changes and the advance and retreat movements which they induce are too rapid. What little we find of incipient vicariance is limited to subspecies or more rarely to species levels. However, the very dynamism of the biogeographic happenings in the Levant and the many large-scale readjustments between local and invading populations are the principal asset of this province. The possibility to study biogeographical processes in action is the most important legacy of Tethys; at least in our egocentric way of thinking.

In an area in which tinkering with the aquatic environments is almost as old as civilization, the study of the biogeographical history is extremely rewarding: in fact, many of the manmade changes are "déjà vu". The Dead Sea shrank many times in the past because of natural causes. So did the flow of the Nile. The close kins of the last tropical fish species which left the inhospitable Mediterranean were among the first to return to the manmade Tethys of Ferdinand de Lesseps. It is said that he who does not learn from history has to repeat it. Since environmental engineering leads often to unwitting duplication of biogeographical history we would be better advised to learn this history beforehand.

Bibliography

Aartsen van, J.J., 1983. Two more Red Sea species recorded for the first time from the Mediterranean coast of Israel. Boll. Malacologico 1–4:37–40.

Aartsen van, J.J. & F. Carrozza, 1979. *Chrysallida fischeri* (Hornung and Mermod, 1925): A Red Sea Species found at the Israeli Mediterranean coast. Boll. Malacologico 15:29–30.

Achituv, Y. & U.N. Safriel, 1980. A new *Chthamalus* (Crustacea, Cirripedia) from intertidal rocks of the Red Sea. Israel J. Zool. 29:99–109.

Adamson, D.A., F. Gasse, F.A. Street & M.A.J. Williams, 1980. Late Quaternary history of the Nile. Nature, Lond. 338:50–55.

Agur, Z. & U.N. Safriel, 1981. Why is the Mediterranean more readily colonized than the Red Sea by organisms using the Suez Canal as a passageway? Oecologia (Berl.) 49:359–361.

Aharoni, I., 1942. Change of dwelling places of some birds in Palestine and an experiment on the reasoning power of *Cinnyris osea* Bonaparte. Bull. zool. Soc. Egypt 4:13–19.

Aleem, A.A., 1969. Marine resources of the United Arab Republic. Stud. Rev. gen. Fish Coun. Mediterr. No.43:1–22.

Aleem, A.A., 1984. The Suez Canal as a Habitat and Pathway for Marine Algae and Seagrasses. In Angel, M.V. (ed.), Marine Science of the Indian Ocean and adjacent waters. Proceedings of the Mabahiss John Murray Symposium. Egypt, 3–5 September 1983, pp. 901–918. Pergamon Press.

Almaca, C., 1985. Evolutionary and Zoogeographical Remarks on the Mediterranean Fauna of Brachyuran Crabs. In Moraitou-Apostolopoulou, M. & V. Kiortsis (eds.), Mediterranean Marine Ecosystems, pp. 347–366. Plenum Press.

Almeida Prado-Por, M.S., 1980. Mysidacea from the Gulf of Elat (Gulf of Aqaba). Israel J. Zool. 29:188–191.

Almeida Prado-Por, M.S., 1983. The diversity and dynamics of Calanoida (Copepoda) in the Northern Gulf of Elat (Aqaba), Red Sea. Oceanol. Acta 6(2):139–145.

Almeida Prado-Por, M.S., 1981. Two new subspecies of the *Diamysis bahirensis* species group (Crustacea Mysidacea) from extreme salinity environmets on the Israel and Sinai coasts. Israel J. Zool. 30:161–175.

Almeida Prado-Por, M.S., R. Ortal & F.D. Por, 1981. *Diamysis* from the brackish river Nahal Taninim in Israel and its associated fauna. Rapp. P.-v. Réun. Commn int. Explor. scient. Mer Méditerr. 27(40):181-182.

Almogi-Labin, A., 1982. Stratigraphic and paleoceanographic significance of late Quaternary pteropods from deep-sea cores in the Gulf of Aqaba (Elat) and northernmost Red Sea. Mar. Micropaleont. 7:53–72.

Almogi-Labin, A., B. Luz & J.C. Duplessy, 1986. Quaternary paleo-oceanography, pteropod preservation and stable isotope record of the Red Sea. Palaeogr. Palaeoclimat. Palaeoecol. 57:195–211.

Alouf, N.J., 1982. Répartition des gammarides d'eau douce au Liban. Polskie Archiwm Hydrobiol. 29(2):247–253.

Alouf, N., 1983. Contribution à la connaissance des cours d'eau du Liban: la zonation biologique du Nahr Qab Ilias. Annls Limnol. 19(2):121–127.

Alouf, N.J., 1984. Cycle de *Marthamea beraudi* Navas dans un cours d'eau du Liban (Plecoptera). Annls Limnol. 20 (1–2):11–16.

Amittai, A. & G. Raz, 1969. Preliminary Survey of the Red Sea Coast of Sinai. Joint Research Project Biota of the Red Sea and the Eastern Mediterranean. Interim Report.

Hebrew University Jerusalem (mimeographed).

Andrea, H.G., 1981. Lysianassidae aus dem Abyssal des Roten Meeres. Bearbeitung der Köderfange von FS "Sonne" - Meseda I (1977) (Crustacea: Amphipoda: Gammaridea). Senckenberg. biol. 61:429–449.

Aron, W. & R.H. Goodyear, 1969. Fishes collected during a midwater trawling survey of the Gulf of Elat and the Red Sea. Israel J. Zool. 18:237–244.

Bacescu, M., 1985. The effects of the geological and physicochemical factors on the distribution of marine plants and animals in the Mediterranean. In Moraitou-Apostolopulou, M. & V. Kiortsis (eds.), Mediterranean Marine Ecosystems, pp. 195-212. Plenum Press.

Baird, W., 1859. Description of several species of Entomostraca from Jerusalem. Ann. Mag. nat. Hist. (Ser. 3) 4:280–283.

Balss, H., 1915. Expeditionen S.M. Schiff "Pola" in das Rote Meer nördliche und südliche Hälfte. 1895/96–1987/98. Zoologische Ergebnisse. XXX. Die Decapoden des Roten Meeres.-I. Die Macruren. Denkschr. Akad. Wiss., Wien 91:1–38.

Balss, H., 1929. Expedition S.M. Schiff "Pola" in das Rote Meer nördliche und südliche Hälfte, 1895/96–1897/98. Zoologische Ergebnisse. XXXVI. Decapoden des Roten Meeres.-IV. Oxyrhyncha und Schlussbetrachtungen. Denkschr. Akad. Wiss., Wien 102:1–30.

Banarescu, P., 1970. Principii si probleme de zoogeografie. Ed. Acad. Republ. Soc. Romania, 260 pp.

Banarescu, P., 1977. Position zoogéographique de l'ichthyofaune d'eau douce de l'Asie Occidentale. Cybium, 3e serie (2):35–55.

Banarescu, P.M., T.T. Nalbant & M. Goren, 1982. The noemacheiline loaches from Israel (Pisces, Cobitidae, Noemachilinae). Israel J. Zool. 31:1–25.

Baranes, A., 1982. Impact of vertical distribution on the biology of two Red Sea sharks. Abstract, IVth European Congress Ichthyology: 18.

Baranes, A. & A. Ben Tuvia, 1979. Two rare carcharinids, *Hemipristis elongatus* and *Iago omanensis*, from the Northern Red Sea. Israel J. Zool. 28:39–50.

Barash Al. & Z. Danin, 1977. Additions to the knowledge of the Indo-Pacific Molluscs in the Mediterranean. Conchiglie, Milano 13:85–116.

Barash, Al., Z. Danin & I. Yaron, 1984. The Genus *Rhinoclavis* (Gastropoda: Cerithiidae) in the Red Sea. Annali Mus. civ. Stor. nat. Giacomo Doria 85:95–117.

Barash, Al. & Z. Danin, 1986. Further additions to the knowledge of Indo-Pacific Mollusca in the Mediterranean Sea (Lessepsian migrants). Spixiana 9(2):117–141.

Barash, Al. & Z. Danin, 1987. Notes on the antilessepsian migration of Mediterranean species of Mollusca into the Indo-Pacific region. Gloria Maris, Antwerpen 26(5,6):81–100.

Barrois, Th., 1982. Liste des phyllopodes recueillis en Syrie. Revue biol. Nord France 5:24–39.

Baruch, U., 1986. Climatic changes in the Dead Sea Rift during the later Quaternary as indicated by the palynological research. Rotem, Bull. Israel Plant Inf. Centre 20:135–139 (in Hebrew).

Bebars, M.I., 1981. Exploitation rationelle des pêcheries égyptiennes: application aux pêcheries des sardinelles (*Sardinella aurita* Valenciennes, 1847) de la baie de Salloum, Égypte. Thesis Univ. Montpellier. 354 pp.

Bebars, M.I. & G. Lasserre, 1983. Analyse des captures des pecheries marines et lagunaires d'Égypte de 1962 a 1976, en liaison avec la construction du haut barrage d'Assouan achevé en 1969. Oceanol. Acta 6(4):417–426.

Beckmann, W., 1984. Mesozooplankton distribution on a transect from the Gulf of Aden to the Central Red Sea during the winter monsoon. Oceanol. Acta 7(1):87–102.

Begin, B.Z., 1986. Lisan lake and its climatological meaning. Rotem, Bull. Israel Plant Inf. Centre 20:31–50 (in Hebrew).

Begin, B.Z., A. Ehrlich & Y. Nathan, 1974. Lake Lisan, The Pleistocene precursor of the Dead Sea. Israel Geol. Surv. Bull. 62, 30 pp.

Bellan-Santini, D., 1985. Mediterranean Benthos: Reflections and Problems Raised by a

Classification of the Benthic Assemblages. In Moraitou-Apostolopoulou, M. & V. Kiortsis (eds.), Mediterranean Marine Ecosystems, pp. 19–48. Plenum Press.

Ben Avraham, Z., 1987. Rift propagation along the southern Dead Sea rift (Gulf of Elat), Tectonophysics, 143:192–200.

Ben-David, Z., 1987. Crocodile River: the last unpolluted stream in the coastal plain. Israel Land and Nature 12(4):143–147.

Ben-Eliahu, M.N., Red Sea serpulids in the Eastern Mediterranean. Proc. 2nd Int. Polychaete Conference, Copenhagen, 1986 (in press).

Ben-Tuvia, A., 1976. Fish collections from the Eastern Mediterranean, the Red Sea and inland waters of Israel. Dept. Zool., Zool. Mus. Hebrew Univ. Jerusalem. 32pp.

Ben-Tuvia, A., 1978. Fishes. In Serruya, C. (ed.), Lake Kinneret, pp. 407–430. Dr. W. Junk Publ.

Ben-Tuvia, A., 1981. Man-induced changes in the Freshwater Fauna of Israel. Fish. Mgmt. 12(4): 138–144.

Ben-Tuvia, A., 1983. The Mediterranean Sea, B. Biological Aspects. In Ketchum, B.H. (ed.), Estuaries and Enclosed Seas, Ecosystems of the World, pp. 239–251. Elsevier, Amsterdam.

Ben-Tuvia, A., 1985. The Impact of the Lessepsian (Suez Canal) Fish Migration on the Eastern Mediterranean Ecosystem. In Moraitou- Apostolopoulou, M. & V. Kiortsis (eds.), Mediterranean Marine Ecosystems, pp. 367–375. Plenum Press.

Benayahu, Y. & Y. Loya, 1977. Space partitioning by stony corals, soft corals and benthic algae on the coral reefs of the Northern Gulf of Eilat (Red Sea). Helgoländer wiss. Meeresunters 30:362–382.

Bender, F., 1968. Geologie von Jordanien. Beiträge zur Regionalen Geologie der Erde ed. H.J. Martin, vol. 7. Borntrager Berlin-Stuttgart. 230 pp.

Benson, R.H., 1975. The origin of the psychrosphere as recorded in the changes of deep-sea ostracode assemblages. Lethaia 8:69–83.

Benson, R.H., R.E. Chapman & L.T. Deck, 1985. Evidence of major events in the South Atlantic and worldwide over the past 80 million years. In Hsu, K.L. & H.J. Weisert (eds.), South Atlantic Paleogeography, pp. 337–350. Cambridge Univ. Press.

Berdugo, V. & B. Kimor, 1968. Considerations on the distribution of pelagic Copepoda in the Eastern Mediterranean. Rapp. P.-v. Réun. Commn Int. Explor. scient. Mer Méditerr. 22(9):85–86.

Berg, L.S., 1948–1949. Ryby presnikh vod S.S.S.R. i sopredelnych stran. vols. 1–3- Moskwa, Leningrad.

Berggren, A. A., 1969. Micropaleontological investigations of the Red Sea cores -- summation and synthesis of results. In Degens, E.T. & D.A. Ross (eds.), Hot brines and heavy metal deposits in the Red Sea, pp. 282–298. Springer.

Berman, T., Y. Azov, A. Schneller, P. Walline & T.W. Townsend, 1986. Extent, transparency and phytoplankton distribution of the neritic waters overlaying the Israeli coastal shelf. Oceanol. Acta 9(4).

Berthelemy, C. & A. Dia, 1982. Plecopteres du Liban (Insecta). Annls Limnol. 18(20):191–214.

Beyth, M., 1980. Recent Evolution and Present State of Dead Sea Brines. In Nissenbaum, A. (ed.), Hypersaline Brines and Evaporitic Environments, pp. 155–165. Elsevier.

Bizon, G., 1985. Mediterranean Foraminiferal Changes as Related to Paleoceanography and Paleoclimatology. In Stanley, J.D. & F.-C. Wezel (eds.), Geological Evolution of the Mediterranean Basin, pp. 453–470. Springer.

Bonatti, E., P. Colantoni, B. Della Vedova & M. Taviani, 1984. Geology of the Red Sea transitional region (22îN–25îN). Oceanol. Acta 7(4):385–398.

Botosaneanu, L., 1963. Les Trichopteres du Lac Houle (Israel). Polskie Pismo ent. 33(2):95–99.

Botosaneanu, L., 1973. Au carrefour des régions orientale, éthiopienne et palearctique. Essai de reconstitution de l'histoire de quelques lignées "cool adapted" de Trichopteres. Fragm. ent. 9(2):61–80.

Botosaneanu, L., 1974. Notes descriptives, faunistiques, ecologiques, sur quelques trichopteres du "trio subtroglophile" (Insecta: Trichoptera). Trav. Inst. Speol. "Emile Racovitza", Bucarest 13:61–75.

Botosaneanu, L., 1984. The Trichoptera of the Levant. In Morse, J.C. (ed.), Proc. fourth Int. Symp. Trichoptera, pp. 39–42. Dr. W. Junk Publ.

Botosaneanu, L., Trichoptera of the Levant. Fauna Palaestina. Ed. Israel Acad. Sci. Jerusalem (in press).

Botosaneanu, L. & A. Gasith, 1971. Contributions taxonomiques et ecologiques a la connaissance des Trichopteres (Insecta) d'Israel. Israel J. Zool. 20:89–129.

Bottger, R., 1987. The Vertical Distribution of Micro- and Small Mesozooplankton in the Central Red Sea. Biol. Oceanogr. 4(4):383–402.

Braithwaite, C.J.R., 1982. Patterns of accretion of reefs in the Sudanese Red Sea. Mar. Geol. 46:297–325.

Briggs, J.C., 1974. Marine Zoogeography, McGraw Hill. New York.

Brinkhurst, R.O. & B.G.M. Jamieson, 1971. Aquatic Oligochaeta of the World. Oliver and Boyd Edinburgh. 860 pp.

Bromley, H.J., 1974. Morpho-karyological types of *Dugesia* (Turbellaria, Tricladida) in Israel, and their distribution patterns. Zool. Scripta 3:239–242.

Bromley, H.J., A note on the Plecoptera of Israel. Israel J. Ent. (in press).

Bromley, H.J., A preliminary outline of the freshwater Hirudinea of Israel. Israel J. Zool. (in press).

Bromley, H.J. & F.D. Por, 1975. The metazoan fauna of a sewage-carrying wadi, Nahal Soreq (Judean Hills, Israel). Freshwat. Biol. 5:121–133.

Brown, E.S., 1951. Aquatic and Semi-Aquatic Hemiptera. In: Expedition to South-West Arabia 1937–1938. 1 (No.17):221–274. British Mus. Nat. History.

Butzer, K.W., 1978. The Late Pleistocenic Environental History of the Middle East. In Brice, W.C. (ed.), The Environmental History of the Near and Middle East since the Last Ice Age. Acad. Press.

Butzer, K.W., 1980. Pleistocene history of the Nile Valley in Egypt and Lower Nubia. In Williams, A.J. & H. Faure (eds.), The Sahara and the Nile, pp. 253–280. Balkema, Rotterdam.

Calvert, S. E., 1983. Geochemistry of Pleistocene sapropels and associated sediments from the Eastern Mediterranean. Oceanol. Acta 6(3):255–267.

Campbell, A. C., 1987. Echinoderms of the Red Sea. In Edwards, A.J. & S.M. Head (eds.), Key Environments: Red Sea, pp. 215–232. Pergamon Press.

Carey, W.S., 1987. Tethys and her Forebears, in K.G. McKenzie (ed.). "Shallow Tethys 2", Balkema, Rotterdam, pp. 3–29.

Casanova, J.P., 1985. Les Chetognathes de la Mer Rouge. Remarques morphologiques et biogéographiques. Description de *Sagitta erythraea* n.sp. Rapp. P.-v. Réun. Commn int. Explor. scient. Mer Méditerr. 29(9):269–274.

Casanova, J.P., 1986. Similarity of Plankton Distribution patterns in two nearly Land-Locked Seas: the Mediterranean and the Red Sea. In Pierrot-Bults, S. van der Spoel, B.J. Zahuranec & R.K. Johnson (eds.), Pelagic Biogeography. UNESCO Technical Papers in Marine Science, 49. 42–46.

Cernosvitov, L. 1938. The Oligochaeta. In Washbourn R. & R.F. Jones "Reports of the Percy Sladen Expedition to the Lake Huleh". Ann. Mag. nat. Hist. 11:535-550.

Chalifa, Y., 1985. *Saurorhamphus judaensis* (Salmoniformes: Enchodontoidei), a new longirostrine fish from the Cretaceous (Cenomanian) of Ein-Yabrud, near Jerusalem. J. Vert. paleont. 5:181–193.

Chappuis, P.A., 1955. Mission Henri Coiffait au Liban (1951). 9. Crustaces d'eau douce cavernicole. Archs Zool. exp. gén. 91:533–536.

Cherbonnier, G., 1986. Holothuries de Mediteranee et du nord de la mer Rouge. Bull. Mus. natn. Hist. nat., Paris, 4-eme serie 8:43–46.

Chervinski, J., 1987. *Tilapia zillii* (Gervais) in the Mediterranean. Bamidgeh 39(4):133–134.

Christiaens J., 1987. Red Sea Malacology. IV. Revision of the Limpets of the Red Sea. Gloria Maris, Antwerpen 26(2)(3):17–54.

CLIMAP project members, 1976. The surface of the Ice-Age Earth. Science, N.Y.

191:1131–1144.
Clark, A.M. & F.W.E. Rowe, 1976. Monograph of shallow-water Indo-West Pacific Echinoderms. British Mus. nat. Hist. 238 pp.
Coad, B.W., 1987. Zoogeography of the Freshwater Fishes of Iran. In Krupp, F., W. Schneider & R. Kinzelbach (eds.), Proceedings of the Symposium on the Fauna and Zoogeography of the Middle East. Beihefte TAVO A 28:213–228.
Cowgill, U., 1969. The waters of Merom: A study of Lake Huleh. 1. Introduction and stratigraphy of a 54 m core. Arch. Hydrobiol. 66:249–272.
Crosskey, R.W., 1967. A preliminary revision of the black-flies (Diptera:Simuliidae) of the Middle East. Trans. R. ent. Soc. Lond. 119(19):1–45.
Crossland, C., 1939. Narrative and List of Stations. Reports on the Preliminary Expedition for the Exploration of the Red Sea in the R.S.S. "Mabahith". Publs mar. biol. Stn Ghardaqa 1: 3–11.
Darom, D. (Masry), 1974. A Systematic and Ecological Study of Some Interstitial Crustacea from Sandy Beaches along the Gulf of Elat (Aqaba). Thesis, Hebrew Universtity of Jerusalem (english summary). 100 pp.
Degens, E.T. & D.A. Ross, 1969. Hot brines and recent heavy metal deposits in the Red Sea. Springer.
Demoulin, G., 1973. Contribution a l'étude des ephemeropteres d'Israel. Introduction et I. Heptagenidae. Bull. Inst. r. Sci. nat. Belg. (Entomologie): 49(8):1-19.
Derin, B. & Z. Reiss, 1973. Revision of marine Neogene stratigraphy in Israel. Israel J. Earth Sci. 22:197–210.
Deuser, W.G. & E.T. Degens, 1969. O^{18}/O^{16} and C^{13}/C^{12} ratios of fossils from the hot-brine deep areas of the Central Red Sea. In Degens, E.T. & D.A. Ross (eds.), Hot brines and recent heavy metal deposits in the Red Sea, pp. 336–374. Springer.
Dexter, D.M., 1988. Sandy beach fauna of Mediterranean and Red Sea coastlines of Israel and of the Sinai Peninsula. Israel J. Zool. 34(3–4):125–138.
Dia, A., 1983. Recherches sur l'écologie et la biogéographie des cours d'eau du Liban Meridional. Thesis. Fac. de Sciences St. Jerome (Aix-Marseille III). 302 pp.
Dia, A. & L. Botosaneanu, 1983. Six éspèces nouvelles de Trichopteres du Liban. Bull. zool. Mus. Amsterdam 9(4):125–135.
Dimentman, Ch. 1976. Distribution and Biology of Anostraca (Crustacea) in Temporary Pools In Israel and Factors Affecting it. Thesis Hebrew University (english summary). 140 pp. + 6 pp. (abstract).
Dimentman, Ch., 1980. Distribution and biology of *Artemia salina* (L) (Crustacea:Anostraca) in Israel and the Sinai Peninsula. Israel J. Zool. 29:199.
Dimentman, Ch., 1981. The rainpool ecosystems of Israel: Geographical distribution of freshwater Anostraca (Crustacea). Israel J. Zool. 30:1–15.
Dimentman, Ch., H.J. Bromley & F.D. Por, Lake Hula: A reconstruction of the fauna of a lost lake. Israel Acad. Sci. (in press).
Dimentman, Ch. & F.D. Por, 1982. Distribution patterns of the inland water fauna in Israel. In Barth, F.G. (ed.), Joint Syposium of the Hebrew University of Jerusalem, Universite Lyon and J.W. Goethe University, Frankfurt 1981, pp. 106–110.
Dimentman, Ch. & F.D. Por, 1985. The distribution of the freshwater Diaptomidae (Copepoda: Calanoida) of Israel and Sinai. Hydrobiologia 127:89–95.
Dowidar, N.M., 1984. Phytoplankton and primary production of the South Eastern Mediterranean. Deep Sea Res. 31(A):983-1000.
Dowidar, N.M. & A.M. Al-Maghrabi, 1971. Observations on the neritic zooplankton in the Abu Qir Bay during the flood season. Rapp. P.-v. Réun. Commn int. Explor. scient. Mer Méditerr. 20(3):385-389.
Dubertret L. & J. Weulersse, 1940. Manuel de Géographie. Syrie, Liban et Proche Orient. Première Partie, La Peninsule Arabique. Imprimerie Catholique Beyrouth. 192 pp.
Dumont, H.J., 1973. The genus *Pseudagrion* Selys in Israel and Egypt, with a key to the regional species (Insecta, Odonata). Israel J. Zool. 22:169-195.
Dumont, H.J., 1974. The genus *Pseudagrion* Selys in Israel and Egypt, with a key to the regional species (Insecta: Odonata). Israel J. Zool. 22:169–195.

Dumont, H.J., 1975. Endemic dragonflies of late Pleistocene age of the Hula Lake area (northern Israel) with notes on the Calopterygidae of the rivers Jordan (Israel, Jordan) and Litani (The Lebanon) and description of *Urothemis edwardsi hulae* subsp. nov. (Libellulidae) Odontologia 4:1-9.

Dumont, H.J., 1979. Limnologie van Sahara en Sahel. (Mimeographed). University of Gent, Belgium. 557 pp.

Dumont, H.J., Odonata. Fauna Palaestina. Israel Acad. Sci. Hum. Jerusalem (in press).

Edwards, F.J., 1987. Climate and Oceanography. In Edwards, A.J. & S.M. Head (eds.), Key Environments: Red Sea, pp. 45–68. Pergamon Press.

Ehrlich, A., 1973. Quaternary diatoms of the Hula Basin (Northern Israel). Israel Geol. Surv. Bull. 58. 38 pp.

Ehrlich, A., 1983. Diatoms from River Dan. In Annonymous, "Land of Streams". Israel Nature Reserves Author. and Nature Protect. Soc. 52 (in Hebrew).

Ehrlich, A., 1985. The eco-biostratigraphic significance of the fossil diatoms of Lake Kinneret. GSI Current Research vol.5 Jerusalem: 24–30.

Ehrlich, A. & I. Dor, 1985. Photosynthetic Microorganisms of the Gavish Sabkha. In G.M. Friedman and W.E. Krumbein (eds.), Hypersaline Ecosystems. The Gavish Sabkha, pp.296–321. Springer.

Ekman, S. 1967. Zoogeography of the Sea. Sidgwick & Jackson, London. 417 pp.

El-Hehyawi, M.L.E., 1974. Changes in salinity and landings of six fish species in the shelf, north to the Nile Delta. Bull Inst. océanogr. Fish. Cairo 4:431–458.

El-Maghraby, A.M. & Dowidar, N.M., 1973. Observations on the zooplankton community in the Egyptian Mediterranean waters. Rapp. P.-v. Réun. Commn int. Explor. scient. Mer Méditerr. 21:527.

El-Sabh, M.I., 1969. Seasonal hydrographic variations in the Suez Canal after the completion of the Aswan High Dam. Kieler Meeresforsch. 25(1):1–18.

Elsayed, A.I.W., O. Guelgorget, G.F. Frisoni, J.M. Rouchy, A. Maurin & J.P. Perthuisot, 1985. Expressions hydrochimiques, biologiques et sédimentologiques des gradients de confinement dans la lagune de Guemsah (Golfe de Suez, Egypte). Oceanol. Acta 8(3):303–320.

El-Sharkavy, S.M. & S.H. Sharaf El-Din, 1983. Great Bitter Lake as a Barrier between the Mediterranean and Red Sea Flows. In Abdel Latif, A.F., A. Refai Bayoumi & M-F. Thompson (eds.), Marine Science in the Red Sea. Bull. Inst. océanogr. Fish. Cairo 9:58–68.

Elster, H.J. & R.A. Vollenweider, 1961. Beiträge zur Limnologie Ägyptens. Arch. Hydrobiol. 57:241–343.

El-Wakeel S., 1984. The Development of Marine Science in Egypt. In Angel, M.V. (ed.), Marine Science in the North-West Indian Ocean and Adjacent Waters. Proceedings of the Mabahiss John Murray International Symposium, Egypt, 3–6 Sept. 1983, pp. 617–638. Pergamon Press.

Emery, K.O. & Y. Bentor, 1960. The Continental Shelf of Israel. Geol. Surv. Israel Bull. 26:25–41.

Eren, J. & I. Dor, 1983. Micro-algae in River Dan. In Annonymous "Land of Streams". Israel Nature Res. Author. and Nature Protect. Soc.:53-60 (in Hebrew).

Esteban, M., 1979–80. Significance of the Upper Miocene coral reefs of the Western Mediterranean. Palaeogeogr. Palaeoclimat. Palaeoecol. 29:169–188.

Evans, G., J.W. Murray, H.E.J. Briggs, R. Bate & P.R. Bush, 1973. The oceanography, ecology, sedimentology and geomorphology of parts of the Trucial Coast Barrier Island complex. In Purser, B.H. (ed.), The Persian Gulf, pp. 233–277. Springer.

Ezzat, A. 1972. The bottom fauna of Lake Edku (Egypt-UAR). Rapp. P.-v. Réun. Commn int. Explor. scient. Mer Méditerr. 20(4):503–505.

Fabricius, F.K., K. Braune, G. Funk. W. Hieke & J. Schmolim, 1985. Plio-Quaternary Sedimentation and Tectonics in the Ionian Area: Clues to Recent Evolution of the Mediterranean. In Stanley, J.D. & F.-C. Wezel (eds.), Geological Evolution of the Mediterranean Basin, pp. 293–306. Springer.

Fairbridge, E. W., 1972. Quaternary Sedimentation in the Mediterranean Region Controlled

by Tectonics, Paleoclimates and Sea Level. In Stanley, D.J. (ed.), The Mediterranean, a Natural Sedimentation Laboratory, pp. 99–113.
Farrand, W.P., 1971. Late Quaternary Paleoclimates of the Eastern Mediterranean Area. In: K.K. Turekian (ed.). "The Late Cenozoic glacial ages." Yale University Press, pp. 529–564.
Fishelson, L., 1977. Stability and instability of marine ecosystems, illustrated by examples from the Red Sea. Helgoländer wiss. Meeresunters 30:18–29.
Fleminger, A. & K. Hulsemann, 1987. Geographical Variation in *Calanus helgolandicus* s.l. (Copepoda, Calanoida) and Evidence of Recent Speciation in the Black Sea Population. Biol. oceanogr. 5:43–81.
Fouda, M.M. & A.M. Hellal, 1987. The Echinoderms of the North-western Red Sea. Asteroidea. Fauna and Flora of Egypt, vol. 2, 71 pp.
Fredj, G., 1974. Stockage et exploitation de données en écologie marine. C - Consideration biogéographiques sur le peuplement benthique de la Mediterranée. Mém. Inst. océanogr. Monaco 7:88 p.
Fredj, G. & L. Laubier, 1985. The Deep Mediterranean Benthos. In Moraitou-Apostolopoulu, M. & V. Kiortsis (eds.), Mediterranean Marine Ecosystems, pp. 109–145. Plenum Press.
Fricke, H. & L. Hottinger, 1983. Coral bioherms below the photic zone in the Red Sea. Mar. Ecol. Prog. Ser. 11:113–117.
Fricke, H. W. & H. Schumacher, 1983. The Depth Limits of Red Sea Stony Corals: An Ecophysiological Problem (A Deep Diving Survey by Submersible). PSZNI Mar. Ecol. 4(20):163–194.
Friedman, G.M., 1985. Gulf of Elat (Aqaba). Geological and Sedimentological Framework. In Friedman, G.M. & W.E. Krumbein (eds.), Hypersaline Ecosystems. The Gavish Sabkha, pp. 39–71. Springer.
Friedman, G.M. & W.E. Krumbein, 1985. Hypersaline Ecosystems. The Gavish Sabkha. Springer. 484 pp.
Friedman, G.M., A. Sneh & R.W. Owen, 1985. The Ras Muhammad Pool: Implicatons for the Gavish Sabkha. In Friedman, G.M. & W.E. Krumbein (eds.), Hypersaline Ecosystems. The Gavish Sabkha, pp.218-237. Springer.
Fryer, G., 1964. Studies on the functional morphology and feeding mechanism of *Monodella argentarii* Stella (Crustacea: Thermosbaenacea). Trans. R. Soc. Edinb. 66:49–90.
Fuchs, Th., 1878. Die geologische Beschaffenheit der Landenge von Suez. Denkschr. Akad. Wiss., Wien 38:25–42.
Furnestin, M.L., 1979. Aspects of the Zoogeography of the Mediterranean plankton. In Spoel van der, S. & A.C. Pierrot Bults (eds.), Zoogeography and Diverstity of Plankton, pp.191–253. Edward Arnold, London.
Gaillard, J.M., 1987. Gasteropodes. In Fischer, W., M.Schneider & M.L. Bauchot (eds.), Fiches FAO d'identification des éspèces pour les besoins de la pêche. Mediterranée et Mer Noire vol.1. Végétaux et Invertebrés. CEE-FAO, Rome pp.513–632.
Galil, B., 1986. Red Sea Decapods along the Mediterranean Coast of Israel: Ecology and Distribution. In Dubinsky, Z. & Y. Steinberg (eds.), Environmental Quality and Ecosystem Stability. Vol. III A/B, pp. 179–183. Bar Ilan Univ. Press Ramat-Gan, Israel.
Galil, B.S., D. Golani & M. Tom, Two new migrant decapods from the Eastern Mediterranean. Crustaceana (in press).
Galil, B. & CH. Lewinsohn, 1981. Macrobenthic Communitie of the Eastern Mediterranean Continental Shelf. PSZNI Mar. Ecol. 2(4):343–352.
Garfunkel, Z., 1981. Internal structure of the Dead Sea leaky transform (rift) in relation to plate kinematics. Tectonophysics 80:81–108.
Garfunkel, Z., 1986. The geology of the Dead Sea Rift. Rotem, Bull. Israel Plant Inf. Centre 20:8–30 (in Hebrew).
Garfunkel Z. & Y. Bartov, 1977. The Tectonics of the Suez Rift. Geol. Surv. Israel Bull. 71. 44 p.
Gasse, F. & J. Delibrias, 1977. Evolution of Lake Abhe (Ethiopia and TFAI) from 70,000 B.P. Nature, Lond. 265:42–45.

Gat, J.R. & M. Magaritz, 1980. Climatic Variations in the Eastern Mediterranean Sea Area. Naturwissenschaften 67:80–87.

Gat, J.R., E. Mazor & Y. Tzur, 1969. The stable isotope composition of the mineral waters in the Jordan Rift Valley, Israel. Hydrology 7:334-352.

Gaudy, R., 1985. Features and Pecularities of Zooplankton Communities from the Western Mediterranean. In Moraitou-Apostolopoulou, M. & V. Kiortsis (eds.), Mediterranean Marine Ecosystems, pp. 279–301. Plenum Press.

Gavish, E., W.E. Krumbein & J. Halevy, 1985. Geomorphology, Mineralogy and Groundwater Chemistry as Factors of the Hydrodynamic System of the Gavish Sabkha. In Friedman, G.M. & W.E. Krumbein (eds.), Hypersaline Ecosystems. The Gavish Sabkha, pp. 186–217. Springer.

George, R.Y. & Menzies, R.J., 1968. Additions to the Mediterranean Deep-Sea isopod fauna (Vema - 14). Rev. roum. Biol. Zool. 13:367.

Gerdes, G., J. Spira & Ch. Dimentman, 1985. The Fauna of the Gavish Sabkha and the Solar Lake - a Comparative Study. In Friedman, G.M. & W.E. Krumbein (eds.), Hypersaline Ecosystems. The Gavish Sabkha, pp. 322–345. Springer.

Gilat-Gottlieb, E., 1959. Study of the Benthos in Haifa Bay. Ecology and Zoogeography of Invertebrates. Spec. Publ. Sea Fish. Res. Stn Haifa. 131 pp. (mimeo. hebrew).

Gilat, E., 1964. The Macrobenthic Invertebrate Communities on the Mediterranean Continental Shelf of Israel. Bull. Inst. océanogr. Monaco 62(1290). 46 pp.

Gilat, E., 1974. Macrobenthic Communities off the Sinai Peninsula in the Mediterranean. Proc. Fifth Sci. Conf. Israel Ecol. Soc. Tel-Aviv, 8–9 May, 1974, B.2 - B.

Girdler, R.W., 1984. The Evolution of the Gulf of Aden and Red Sea. In Angel, M.V. (ed.), Marine Science of the North-West Indian Ocean and Adjacent Water. Proceedings of the Mabahiss John Murray International Symposium, Egypt, 3–6 September 1983. pp.747–762. Pergamon Press Oxford.

Godeaux, J., 1983. Les Doliolides de la Mer Rouge. Rapp. P.-v. Réun. Comm. int. Explor. Mer Méditerr. 28(9).

Golani, D., 1987. Comparison of morphomeristical variations of Mediterranean and Red Sea populations of the Suez Canal migrant *Sargocentron rubrum*. Centro 3:25–31.

Golani, D., 1987. The Red Sea pufferfish *Torquigener flavimaculosus* a new Suez Canal migrant in the Eastern Mediterranean (Pisces:Tetraodontidae). Senckenberg. mar. 19(5–6):339–343.

Golani, D., 1988. Aspects of colonization of two Red Sea species, the goldband goatfish *(Upeneus moluccensis)* and the brownband goatfish (*U. asymmetricus*), migrants in the Mediterranean sea. Thesis Hebrew University (in Hebrew with English summary). 124 + V pp.

Golani, D. & A. Ben-Tuvia, 1986. New records of fishes from the Mediterranean coast of Israel, including Red Sea immigrants. Cybium 10(3):285–291.

Goren, M., 1974. The Freshwater Fishes of Israel. Israel J. Zool. 23:67–118.

Gorgy, S., 1966. Les pêcheries et le milieu marin dans le secteur mediterranéen de la République Arabe Unie. Revue Trav. Inst. (scient. tech.) Pêch. marit. 30(1):25–80.

Gottlieb, E., 1959. A Study of the Benthos in Haifa Bay: Ecology and Zoogeography of Invertebrates. Thesis, Hebrew University of Jerusalem (english summary).

Grecchi G., 1978. Problems connected with the recorded occurence of some mollusks of Indopacific affinity in the Pliocene of the Mediterranean area. Riv. ital. Paleont. Stratigr. 84:797–812.

Gruvel, A., 1931. Les États de Syrie. Richesses marines et fluviales. Exploitation actuelle, avenir. Soc. Ed. Geogr. Marit. Colon. Franc. Paris. 453 pp.

Guergues, S.K. & Y. Halim, 1973. Chetognathes du plancton d'Alexandria. II. Un specimen mur de *Sagitta neglecta* Aida en Mediterranée. Rapp. P.-v. Réun. Commn int. Explor. scient. Mer Méditerr. 21:497.

Guerre, A. & P. Sanlaville, 1970. Sur les hauts niveaux marins quaternaires du Liban. Hannon 5:21–27.

Gvirtzman, G. & Buchbinder B., 1977. The dessication events in the Eastern Mediterranean as compared with other Miocene dessication events in the basins around the Mediter-

ranean. In Bijou-Duval, B. & L. Montadert (eds.), The Structural History of the Mediterranean Basins, pp. 411–420. Ed. Techniques. Paris.
Haas, G., 1953. On the occurence of Hippopotamus in the Iron Age of the Coastal Area of Israel (Tel Qasileh). Bull. Am. Sch. Orient. Res. 132:30–34.
Haas, G., 1978. A Cretaceous Pleurodire turtle from the surroundings of Jerusalem. Israel J. Zool. 27:20–33.
Halim, Y., 1969. Plankton of the Red Sea. Oceanogr. mar. biol. A. Rev. 7:231–274.
Halim, Y., 1976. Marine biological studies in Egyptian Mediterranean waters: a Review. Acta Adriat. 18(2):29–38.
Halim, Y., 1984. Plankton of the Red Sea and the Arabian Gulf. In Angel, M.V. (ed.), Marine Science of the North-West Indian Ocean and Adjacent Waters. Proceedings of the Mabahiss John Murray International Symposium., Egypt, 3–6 September 1983, pp. 969–982. Pergamon Press.
Halim, Y, H.A. Sultan & A. Saaman, 1983. Effect of the Asswan High Dam on the phytoplankton standing crop around Alexandria. Rapp. P.-v. Réun. Commn int. Explor. scient. Mer Méditerr. 28:9, 63–64.
Hammerton, D., 1972. The Nile River. In: River Ecology and Man, Acad. Press, p. 171–215.
Harding, J.P., 1938. Crustacea, Copepoda. In Washbourne, R. & R.F. Jones (eds.), Report of the Percy Sladen Expedition to Lake Huleh: a Contribution to the Fresh Waters of Palestine. Ann. Mag. nat. Hist. 2(11):551–552.
Hartland-Rowe, R., 1967. An annotated catalogue of phyllopod Crustacea recorded from Israel, with a key for their identification. Israel J. Zool. 16:88–95.
Head, S.M., 1987. Introduction. In Edwards, A.J. & S.M. Head (eds.), Key Environments: Red Sea, pp. 1–21. Pergamon Press.
Head, S.M., 1987. Corals and Coral Reefs of the Red Sea. In Edwards, A.J. & S.M. Head (eds.), Key Environments: Red Sea, pp. 128–151. Pergamon Press.
Heckel, J.J., 1843. Abbildungen und Beschreibungen der Fische Syriens, nebst einer neuen Klassification und Characteristik sämmtlicher Gattungen der Cypriniden. Stuttgart, 258 pp.
Hedgpeth, J.W., 1957. Estuaries and Lagoons. II. Biological Aspects. In Hedgpeth, J.W. (ed.), Treatise on Marine Ecology and Palaeoecology. Geol. Soc. Am. Mem. 67:49–54.
Herbst, G.N., 1982. New records of *Typhlocirolana* species (Crustacea:Isopoda) from Israel: ecological and biogeographical significance. Israel J. Zool. 31:47–53.
Herbst, G.N. & S.R. Reice, 1982. Comparative leaf litter decomposition in temporary and permanent streams in semi-arid regions of Israel. J. Arid Envir. 5:305–318.
Herbst, G.N. & Ch. Dimentman, 1983. Distributional patterns and habitat characteristics of Amphipoda (Crustacea) in the inland waters of Israel and Sinai. Hydrobiologia 98:17-24.
Herbst, G.N. & H.J. Bromley, 1984. Relationship between habitat stability, ionic composition and the distribution of aquatic invertebrates in the desert regions of Israel. Limnol. oceanogr. 29:495–503.
Herbst, G.N. & H.K. Mienis, 1985. Aquatic invertebrate distribution in Nahal Tanninim, Israel. Israel J. Zool. 33:51–62.
Herman, Yvonne, 1981. Distribution and Ecology of Recent Pteropoda and Planktonic Foraminifera in the Mediterranean Sea. Rapp. P.-v. Réun. Commn int. Explor. scient. Mer Méditerr. 27:7, 153–154.
Herrnstadt I. and C.C. Heyn, Bryophyte Flora of Israel. Israel Acad. Sci. Hum. Jerusalem (in press).
Hey, R.W., 1978. Horizontal Quaternary Shorelines in the Mediterranean. Quatern. Res. 10:197–203.
Holthuis, L.B., 1963. On red coloured shrimps (Decapoda, Caridea) from tropical land-locked saltwater pools. Zool. Meded., Leiden 38:261–279.
Holthuis, L.B., 1973. Caridean shrimps found in land-locked saltwater pools of four Indopacific localities (Sinai Peninsula, Funafuti atoll, Maui and Hawaii Islands) with description of a new genus and four new species. Zool. Verh., Leiden 128:1–48.
Holthuis, L.B., 1987. Crevettes, Homards, Langoustines, Langoustes et Cigales. In Fischer, W., M.Schneider & M.L. Bauchot Fiches FAO d'identification des éspèces pour les

besoins de la pêche. Mediterranée et Mer Noire vol. 1. Végétaux et Invertebrés, CEE-FAO, Rome pp.293–319.

Holthuis, L.B. & E. Gottlieb, 1958. An annotated List of the Decapod Crustacea of the Mediterranean Coast of Israel, with an Appendix Listing the Decapoda of the Eastern Mediterranean. Bull. Res. Counc. Israel 7B:1–126.

Hopkins, T.S., 1985. Physics of the Sea. In Margaleff, R. (ed.), Key Environments. Western Mediterranean, pp. 100–125. Pergamon Press, Oxford.

Horowitz, A., 1973. Development of the Hula Basin. Israel J. Earth Sci. 22:107–139.

Horowitz, A., 1978. The Quaternary evolution of the Jordan Valley. In Serruya, C. (ed.), Lake Kinneret, pp. 33–44. Dr. W. Junk Publishers.

Horowitz, A., 1979. The Quaternary of Israel. Academic Press.

Horowitz, A., 1987. Travertines of the Arid Region. Oxygen Isotope Stages, and Late Quaternary Climates in Israel. Quatern. Res. 27:103–105.

Horowitz, A., 1988. The Quaternary environments and paleogeography in Israel. In Yom-Tov, Y. & E. Tchernov (eds.), Zoogeography of Israel, pp. 35–57. Dr. W. Junk Publ.

Hsu K.J., L. Montadert, D. Bernoulli, M.B. Cita, A. Erikson, R.E. Garrison, R.B. Kidd, F. Melieres, C. Muller & R. Wright, 1978. History of the Mediterranean salinity crisis. Initial Reports of DSDP. 42/1. 1053–1078.

Hubault, F., 1937. *Sphaeromicola sphaeromidicola* nov. sp. commensal de *Sphaeromides virei* Valle, en Istrie et consideration sur l'origine de diverses éspèces cavernicoles perimediterranéennes. Archs Zool. exp. gén. 180:11–24.

Hutchinson, G.E. & Cowgill, U.M., 1973. The waters of Merom: A study of Lake Huleh. III. The major chemical constituents of a 54 m core. Arch. Hydrobiol. 77:145–185.

Illies, J. & L. Botosaneanu, 1963. Problèmes et methodes de la classification et de la zonation écologique des eaux courants, considerées surtout du point de vue faunistique. Mitt. Soc. int. Limnologie 12:1–57.

Inbar, M., 1982. Spatial and temporal aspects of man-induced changes in the hydrological and sedimentological regime of the upper Jordan River. Israel J. Earth Sci. 31:53–66.

Issar, A. & Y. Eckstein, 1969. The lacustrine beds of Wadi Feiran, Sinai. Israel J. of Earth Sci. 18(1):29–32.

Issar, A. & H. Bruins, 1983. Special climatic conditions in the Sinai and Negev during the most Upper Pleistocene. Palaeogeogr. Palaeoclimat. Palaeoecol. 43:63–72.

Issar, A. & H. Tsoar, 1987. Who is to blame for the desertification of the Negev ? IAHS Publ. no. 168, 7 pp.

Issar, A., Ch. Tsoar, I. Gilead & A. Zangvil, 1987. A Palecoclimatic Model to Explain Depositional Environments during Late Pleistocene in the Negev. In Berkofsky, L. & M.G. Wurtele (eds.), Progress in Desert Research, pp. 302–309. Rowman and Littlefield.

Jäch, M.A. & J. Margalit, 1987. Distribution of Dystiscids in springs of the Western Dead Sea area (Coleoptera: Dytiscidae). The Coleopts Bull. 41:327–334.

Jäch, M.A. & R. Ortal, 1988. The large water beetles of Israel on the verge of extinction? Shappirit. Tel Aviv 6:37–49 (in Hebrew).

James, D.B. & J.S. Pearse, 1969. Echinoderms from the Gulf of Suez and the northern Red Sea. J. mar. biol. Ass. India 11:78–125.

Jones, D.A., 1984. Crabs of the mangal ecosystem. In Por, F.D. & I. Dor (eds.), Hydrobiology of the Mangal, pp. 89–109. Junk Publishers, The Hague.

Jones, D.A., M. Ghamrawy & M.I. Wahbeh, 1987. Littoral and Shallow Subtidal Environments. In Edwards, A.J. & S.M. Head (eds.), Key Environments: Red Sea, pp. 169–193. Pergamon Press.

Jones, R.F., 1940. Report of the Percy Sladen Expedition to Lake Huleh: a Contribution to the Study of the Fresh Waters of Palestine. I. The Plant Ecology of the District. J. Ecol. 28:357–376.

Kafri, U. & A. Arad. 1978. Paleohydrology and migration of the ground-water divide in regions of tectonic instability in Israel. Geol. Soc. Am. Bull. 89:1723–1732.

Kafri, U., B. Lang, A. Ehrlich, S. Moshkovitz, M. Magaritz & A. Kaufman, 1981. Paleolimnological studies of the Hula Basin: recent advances. Geol. Surv. Israel. Current Research 1980, 48–53.

Kaiser, K., E.K. Kempf, Arl. Leroi-Gourhan & H. Schutt, 1973. Quartärstratigraphische Untersuchungen aus dem Damaskus-Becken und seiner Umgebung. Z. Geomorph. N.F. 17/3:263–355.

Kaplan, I.R. & A. Friedman, 1970. Biological productivity in the Dead Sea. Part I. Microorganisms in the water column. Israel J. Chem. 8:513–528.

Karaman, G.S., 1986. First discovery of genus *Niphargus* Sch. in Iraq, Israel and adjacent regions, with description of *N.itus*, new species (Fam. Niphargidae). (Contrib. knowledge Amphipoda 153) Polyopriveda i Sumarstvo, Titograd 32 (1):13-36.

Karmon, Y., 1960. The Drainage of the Huleh Swamps. Geogrl Rev. 50(2):160–193.

Kassas, M. & M.A. Zahran, 1967. On the ecology of the Red Sea littoral salt marsh, Egypt. Ecol. Monogr. 37:297–315.

Kaufman, A., 1971. U-Series dating of Dead Sea basin carbonates. Geochim. Cosmochim. Acta 35:1269–1281.

Kempf E.K., 1973. Ostrakoden aus dem Jungpleistozän des Damaskus-Beckens. In Kaiser, K., E.K. Kempf, Arl. Leroi-Gourhan & H. Schutt "Quartärstratigraphische Untersuchungen aus dem Damaskus-Becken und seiner Umgebung". Z. Geomorph. N.F. 17:3,336–346.

Khalil, A.N., El-Maghraby, N.M. Dowidar & D.L. El-Zawawy, 1983.Seasonal variations of the Zooplankton Biomass in the Eastern Harbor of Alexandria, Egypt. Rapp. P.-v. Réun. Commn int. Explor. scient. Mer Méditerr. 28:9, 217–218.

Khouri, R.G., 1981. The Jordan Valley. Life and Society below Sea Level. Longman, London, 238 pp.

Kiener, A., 1978. Écologie, physiologie et économie des eaux saumatres. Masson, Paris. 220p.

Kimor, B. & B. Golandsky, 1977. Microplankton of the Gulf of Elat: aspects of seasonal and bathymetric distribution. Mar. Biol. 42:55–67.

Kinsman, J.J., 1964. Reef coral tolerance of high temperatures and salinities. Nature, Lond. 202: 1280–1282.

Kinzelbach, R., 1980. Hydrobiologie am Orontes. Natur. u. Museum 110(1):9‰18.

Kinzelbach, R.K., 1985. *Potamon potamios ghab* n.ssp. aus dem Orontes-System. Senckenberg. biol. 66(1–3):119–122.

Kinzelbach, R., 1987. Faunal History of Some Freshwater Invertebrates of the Northern Levant. In Krupp, F., W. Schneider & R. Kinzelbach (eds.), Proceedings of the Symposium on the Fauna and Zoogeography of the Middle East. Mainz 1985. Beihefte TAVO A 28:41–61.

Kinzelbach, R.K. & B. Koster, 1985. Die Süsswassergarnele *Atyaephyra desmaresti* (Millet, 1832) in den Levante-Landern. Senckenberg biol. 66(1–3):127–134.

Kinzelbach, R., F. Krupp & W. Schneider, 1987. Levant. Freshwater Fauna. Map A VI 12. "Tübinger Atlas des Vorderen Orients (TAVO)". Ludwig Reichert Wiesbaden.

Kinzelbach, R.K. & G. Roth, 1984. Patterns of distribution of some freshwater molluscs of the Levant region. Fol. Hist. nat. Mus. Matr. Budapest 9:115–120 (8 maps).

Kiortsis, V., 1985. Mediterranean Marine Ecosystems: Establishment of Communities in Transitional and partly Isolated Areas. In Mouraitou- Aopostolopoulou, M. & V. Kiortsis (eds.), Mediterranean Marine Ecosystems. pp. 377–385. Plenum Press.

Kisseleva, M.I., 1977. Qualitative distribution of the meiobenthos in the Red Sea. In Kisseleva, M.I. & A.A. Murina (eds.), The Benthos of the Red Sea Shelf. Naukova Dumka, Kiev.

Klausewitz, W., 1968. Remarks on the zoogeographical situation of the Mediterranean and the Red Sea. Annali Mus. civ. Stor. nat. Giacomo Doria 77:323-328.

Klausewitz, W., 1975. Fische aud dem Roten Meer XV. *Cabillus anchialinae*, eine neue Meergründel von der Sinai Halbinsel (Pisces, Gobiidae, Gobiinae). Senckenberg. biol. 56:203–207.

Klausewitz, W., 1983. Die Entwicklung des Roten Meeres und seiner Küstenfische. I. Evolutionszentrum. II. Palaeogeographie, Palaeökologie und Endemitenentwicklung. Natur. u. Museum 113(4):103–111; 113(12):349–368.

Klausewitz, W., 1986. Zoogeographic Analysis of the Vertical Distribution of the Deep Red

Sea Ichthyofauna, with a New Record. Senckenberg. marit. 17(4/6): 279–292.
Klausewitz, W. & H, Fricke, 1985. Fische aus dem Roten Meer. XVI. On the occurence of *Chaetodon jaykari* Norman, in the Deep Water of the Gulf of Aqaba, Red Sea. Senckenberg. mar. 17(1/3):1–13.
Klein, C., 1961. On the fluctuation of the level of the Dead Sea, since the beginning of the 19th Century. State of Israel Minst. Agric. Hydrological Service.
Klein, C., 1982. Morphological evidence of lake level changes on the Western shore of the Dead Sea. Israel J. Earth Sci. 31:67–94.
Klein, M., 1988. The Geomorphology of Israel. In Yom-Tov, Y. & E. Tchernov (eds.), The Zoogeography of Israel, pp. 59–78. Dr. W. Junk Publ.
Klinker, J.,Z. Reiss, C. Kropach, I. Levanon, H. Harpaz, E. Halicz & G. Assaf, 1976. Observations on the circulation pattern in the Gulf of Elat (Aqaba) Red Sea. Israel J. Earth Sci. 25:85–103.
Kocatas, A., 1981. Liste Preliminaire et Repartition des Crustaces decapodes des eaux turques. Rapp. P.-v. Réun. Commn int. Explor. scient. Mer Méditerr. 24/4:161.
Koch, S., 1988. Mayflies of the Northern Levant (Insecta: Ephemeroptera). Zoology in the Middle East, Vol. 2:89–109.
Kosswig, C., 1967. Tethys and its relation to the peri-Mediterranean faunas of fresh-water fishes. In Adams, C.G. & D.V. Ager (eds.), Aspects of Tethyan Biogeography. Vol. 7, pp. 313–324. The Systematic Association London.
Krumbein, W.E. & Y. Cohen, 1974. Biogene, klastische und evaporitische Sedimentation in einem mesothermen, monomiktischen ufernahen See (Golf von Aqaba). Geol. Rdsch. 63:1035–1065.
Krupp, F., 1980. Die Verbreitung syrischer Süsswasserfische unter dem Einfluss des Menschen. Natur. u. Museum 110:157–164.
Krupp, F., 1982. *Garra tibanica ghorensis* subsp.nov. (Pisces:Cyprinidae), an African element in the cyprinid fauna of the Levant. Hydrobiologia 88:319–324.
Krupp, F., 1983. Fishes of Saudi Arabia. Freshwater fishes of Saudi Arabia and adjacent regions of the Arabian Peninsula. Fauna of Saudi Arabia vol. 5:568–636.
Krupp, F., 1987. Freshwater Ichthyogeography of the Levant. In Krupp, F., W. Schneider & R. Kinzelbach (eds.), Proceedings of the Symposium on the Fauna and Zoogeography of the Middle East, Mainz, 1985. Beihefte TAVO A 28:229–237.
Krupp, F. & W. Schneider, 1988. Die Süsswasserfauna des Vorderen Orientes. Anpassungsstrategien und Besiedlungsgeschichte einer zoogeographischen Übergangszone. Natur. u. Volk. 118(17):193–213.
Kugler, J., 1978. Chironomidae and Trichoptera. In Serruya, C. (ed.) Lake Kinneret, pp. 369–376. Dr. W. Junk Publ.
Kugler, J. & H. Chen, 1968. The distribution of Chironomide larvae in Lake Tiberias (Kinneret) and their occurence in the food fish of the lake. Israel J. Zool. 17:97–115.
Kugler, Y. & F. Reiss, 1973. Die *triangularis*-Gruppe der Gattung *Tanytarsus* v.s.W. (Chironomidae, Diptera). Ent. Tidskr. 94:59-82.
Kugler, J. & D. Wool, 1968. Chironomidae (Diptera) from the Hula Nature Preserve, Israel. Ann. zool. Fennici 5:76–83.
Kuhlmann, D.H.H., 1983. Composition and ecology of deep water coral associations. Helgoländer wiss. Meeresunters 36:183-204.
Kuiper, J.G.J., 1981. The distribution of *Pisidium tenuilineatum* Stelfox and *Pisidium annandalei* Prashad in the Mediterranean area. Basteria 45:79–84.
Kukla, G., 1980. End of the last Interglacial: a Predictive Model of the Future? Palaeoecology of Africa vol. 12:395–408.
Kullenberg, B., 1952. On the salinity of the water contained in marine sediments. Meddn oceanogr. Inst. Götenberg 21:1–38.
Laborel, J., 1981. Peuplements fossiles des niveaux marins surélevées holocenes dans l'arc Egéen. J. Etud. Syst. evol. CIEMS Cagliari. p.151.
Lacombe, H. & P. Tchernia, 1960. Quelques traits généraux de hydrologie Mediteranéene après divers campagnes hydrologiques récentes en Mediterranée, dans le prôche Atlantique et dans le détroit de Gibraltar. Cah. océanogr. 12(8):527–548.

Lakkis, S., 1973. Note préliminaire sur la présence et la repartition des copepodes dans les eaux superficielles libanaises. Rapp. P.-v. Réun. Commn int. Explor. scient. Mer Méditerr. 21:459.

Lakkis, S., 1976. Sur la présence dans les eaux Libanaises de quelques Copepodes d'origine indo-pacifique. Rapp. P.-v. Réun Commn int. Explor. scient. Mer Méditerr. 23:9,83–85.

Lakkis, S., 1983. Le plancton des eaux libanaises (Mediterranée Orientale). Caractéristiques biogéographiques. J. Etud. Syst. Biogeogr. CIESM, Cagliari. p. 59.

Larsen, H., 1980. Ecology of Hypersaline Environments. In Nissenbaum, A. (ed.), Hypersaline Brines and Evaporitic Environments, pp. 23–39. Elsevier.

Lattin, G. de, 1967. Grundriss der Zoogeographie. Gustav Fischer Stuttgart. 602 pp.

Laubscher, H. & D. Bernoulli, 1977. Mediterranean and Tethys. In Nairn, A.E.M., W.H. Kates & F.G. Stehli (eds.), The Ocean Basin and Margins. The Eastern Mediterranean. Vol. 4, pp.1–28. Plenum Press.

Le Pichon X. & J. Francheteau, 1978. A Plate-Tectonic Analysis of the Red Sea-Gulf of Aden area. Tectonophysics 46:369–406.

Lernau, H., 1988. Subfossil remains of Nile perch (*Lates* cf. *niloticus*), first evidence from ancient Israel. Israel J. Zool. 34(3–4):225–236.

Lerner-Seggev, R., 1968. The fauna of Ostracoda in Lake Tiberias. Israel J. Zool. 17:117-143.

Leroi-Gourhan, A., 1973. Palynologische Befunde aus Quartärablagerungen im Damaskus Becken und seine Rahmenbereichen. In Kaiser, K., E.K. Kempf, Arl.

Leroi-Gourhan, A. & H. Schütt, Quartärstratigraphische Untersuchungen aus dem Damaskus-Becken und seiner Umgebung. Z. Geomorph. N.F. 17:3, 303–318.

Levanon-Spanier, I., E. Padan & Z. Reiss, 1979. Primary Production in a desert-enclosed sea - the Gulf of Elat (Aqaba) Red Sea. Deep Sea Res. 26:673–685.

Leventer, H., 1984. Biological Control of Reservoirs by Fish. Mekoroth Water Company, Israel. 71 pp.

Levi, Cl., 1957. Spongiaires des côtes d'Israel. Bull. Res. Counc. Israel 6B:201–212.

Lewinsohn, Ch., 1969. Die Anomuren des Roten Meeres (Crustacea, Decapoda: Paguridea, Galatheidea, Hippidea). Zool. Verh., Leiden no.164, 213 pp.

Liebmann, E., 1935. Oceanographic observations on the Palestine coast. Rapp. P.-v. Réun. Commn int. Explor. scient. Mer Méditerr. 9:181–185.

Liere Van, W. J., 1960/61. Observations on the Quaternary of Syria. Ber. Rijksdienst oudheidk. Bodemonderz. 10/11:7–69.

Linnavuori, R., 1960. Hemiptera of Israel I. Ann. zool. Soc. Vanamo. 22(1):1–71.

Linnavuori, R. 1964. Hemiptera of Egypt, with remarks on some species of the adjacent Eremian region. Ann. zool. Fennici 1:306–356.

Lipkin, Y., 1977. Seagrass vegetation of Sinai and Israel. In McRoy, C.P. & C. Helferich (eds.), Seagrass Ecosystems: A Scientific Perspective, pp. 263–293. Dekker, New York.

Lipkin, Y., 1983. Aquatic plants of River Dan. In Annonymous, "Land of the Streams". Israel Nature Reseves Author. and Natyre Protect. Soc. (in Hebrew).

Livermore, R.A., & A.G. Smith, 1985. Some Boundary Condition for the Evolution of the Mediterranean Region. In Stanley, J.D. & F.-C. Wezel (eds.), Geological Evolution of the Mediterranean Basin, pp. 83–98.

Livnat, A. & J. Kronfeld, 1985. Paleoclimatic implications of U-series dates for lake sediments and travertines in the Arava rift valley, Israel. Quatern. Res. 24:164–172.

Livnat, A. & J. Kronfeld, 1987. Reply to comment on "Travertines of the Arid Regions, Oxygen Isotope Stages, and late Quaternary Climate of Israel" by A. Horowitz. Quatern. Res. 27:106–107.

Locard, A., 1883. Malacologie des Lacs de Tiberiade, d'Antioche et d'Homs. Archs Mus. Hist. nat. Lyon 3:195–293.

Löffler, H., 1967. Limnology. In, J.M. Boyd ed.: International Jordan Expedition. IBP/CT Section London:25–39.

Lortet, L., 1883. Études zoologiques sur la faune du lac de Tiberiade suivi d'un aperçu sur la faune des lacs d'Antiochie et de Homs. I. Poissons et reptiles du lac de Tiberiade et de quelques autres parties de la Syrie. Archs Mus. Hist. nat. Lyon 3:99–108.

Lotan, E., 1982. The killifish *Aphanius dispar*. Israel Land and Nature 8:28–30.
Loya, Y., 1972. Community structure and species diversity of hermatypic corals at Eilat, Red Sea. Mar. Biol. 13:100–123.
Loya, Y., 1976. Recolonisation of Red Sea corals affected by natural catastrophies and man-made perturbations. Ecology 57(2):278–289.
Luksch, J. 1898. Expedition S.M. Schiff "Pola" in das Rothe Meer nördliche Hälfte (Oct. 1895 - Mai 1896). Wissenschaftliche Ergebnisse VI. Physikalische Untersuchungen. Denkschr. Akad. Wiss., Wien 65: 351–422.
Luz, B., 1979. Palaeoceanography of the post-glacial Eastern Mediterranean. Nature, Lond. 278:847–848.
Luz, B., L. Heller-Kallai & A. Almogi Labin, 1984. Carbonate mineralogy of late Pleistocene Sediments from the Northern Red Sea. Israel J. Earth Sci. 33:157–166.
Mancy, K.H., 1981. The environmental and ecological impacts of the Aswan High Dam. In Shuval, H. (ed.), Developments in Arid Zone Ecology and Environmental Quality, pp. 83–99. Balaban I.S.S. Philadelphia.
Mangold, K. & S.V. Boletzky, 1987. Cephalopodes (seiches, calmars et poulpes/pieuvres). In Fischer, W., M. Schneider & M.L. Bauchot (eds.), Fiches FAO d'identification des éspèces pour les besoins de la pêche. Mediterranée et Mer Noire, vol. 1 Végétaux et Invertebrés. pp. 633–714 CEE-FAO Rome.
Marenzeller, E. von, 1895. Echinodermen. Expeditionen S.M. Schiff "Pola" Zoologische Ergebnisse. V. Denkschr. Akad. Wiss., Wien.
Margaleff, R., 1985. Introduction to the Mediterranean. In Margaleff, R. (ed.), Key Environments. Western Mediterranean, pp. 1–16. Pergamon Press, Oxford.
Margalit, J. & A.S. Tahori, 1973. The mosquito fauna of Sinai. J. med. Ent. 10:89-96.
Martinez-Ansemil, E. & Giani N., 1987. The distribution of aquatic oligochaets in the South and Eastern Mediterranean area. Hydrobiologia 155:293–303.
Mastaller, M., 1987. Molluscs of the Red Sea. In Edwards, A.J. & S.M. Head (eds.), Key Environments: Red Sea, pp. 194–214. Pergamon Press.
Mazloum, S., 1939. L'Afrine. Étude hydrologique. Revue Géorg. phys. Géol. dyn. 12:37–164; 189–305.
Mazor, E., 1978. Mineral waters of the Kinneret basin and possible origin. In Serruya, C. (ed.), Lake Kinneret, pp.103–119. Dr. W. Junk Publ.
Mazor, E. & F. Mero, 1969. The origin of the Tiberias-Noit mineral water association in the Tiberias-Dead Sea Rift Valley Israel. J. Hydrol. 7: 318–333.
Mazor, E. & M. Molcho, 1972. Geochemical studies on Feskha springs, Dead Sea Basin. J. Hydrol. 18:289-303.
McClure, H.A., 1976. Radiocarbon chronology of Late Quaternary lakes in the Arabian Desert. Nature, Lond. 263:755.
McKenzie, J.A., H.C. Jenkyns & G.G. Bennet, 1979–1980. Stable isotope study of the cyclic diatomite claystones from the Tripoli formation, Sicily: a prelude to the Messinian salinity crisis. Palaeogeogr. Palaeoclimat. Palaeoecol. 29: 125–141.
McKenzie, J.A. & H. Oberhansli, 1985. Paleoceanographic expressions of the Messinian salinity crisis. In Hsu, K.J. & H.J. Weissert (eds.), South Atlantic paleoceanography, pp. 99–123. Cambridge Univ. Press.
McKenzie, K.G., 1987. Tethys and her progeny. In: K.G. McKenzie (ed.). "Shallow Tethys 2", Balkema, Rotterdam, pp. 501–523.
Menzies, R., 1972. Biological history of the Mediterranean Sea with reference to the abyssal benthos. Rapp. P.-v. Réun. Commn int. Explor. scient. Mer. Méditerr. 207:209.
Mergner, H., 1971. Structure, ecology and zonation of Red Sea Reefs (In comparison with South Indian and Jamaican reefs). Symp. zool. soc. Lond. 28:141-161.
Mergner, H., 1984. The Ecological Research on coral reefs of the Red Sea. In Angel, M.V. (ed.), Marine Science of the North-West Indian Ocean and adjacent waters. Proceedings of the Mabahiss John Murray International Symposium, Egypt, 3–6 September 1983, pp. 855–884. Pergamon Press.
Mergner, H. & H. Schumacher, 1981. Quantitative Analyse der Korallenbesiedlung eines Vorrifareales bei Aqaba (Rotes Meer) Helgoländer wiss. Meeresunters 34:337-354.

Mergner, H. & A. Svoboda, 1977. Productivity and seasonal changes in selected reef areas in the Gulf of Aqaba (Red Sea). Helgoländer wiss. Meeresunters. 30:383–399.

Mero, F., 1978. Hydrology. In Serruya, C. (ed.), Lake Kinneret, pp. 87–102. Dr. W. Junk Publ.

Meulenkamp, J.E., 1985. Aspects of the Late Cenozoic Evolution of the Aegean Region. In Stanley, J.D. & F.-C. Wezel (eds.), Geological Evolution of the Mediterranean basin, pp. 307–321. Springer.

Miller, A.G. & R.G. Munns., 1974. The Bitter Lake salt. L'Océanographie Physique de la Mer Rouge. Symposium. Assoc. Intern. Sci. Phys. et Oceanol. CNEXO. Actes Coll. 2:295-309.

Mor, D. & G. Steinitz, 1983. K-Ar age determination of the Cover Basalt surrounding the Sea of Galilee. Israel Geol. Soc. Annual Meeting. pp.62–64.

Mor, D. & G. Steinitz, 1985. K-Ar age of the Neogene-Quaternary basalts around the Yarmouk Valley. Israel Geol. Soc. Annual Meeting. pp. 77–78.

Moraitou-Apostolopoulou, M., 1985. The Zooplankton Communities of the Eastern Mediterranean (Levantine Basin, Aegean Sea): Influences of Man-Made Factors. In Moraitou-Apostolopoulou, M. & V. Kiortsis (eds.), Mediterranean Marine Ecosystems, pp. 303–331. Plenum Press.

Morcos, S.A., 1967. Effect of the Aswan High Dam on the current regime in the Suez Canal. Nature, Lond. 214(5091):901–902.

Morcos, S.A., 1970. Physical and Chemical Oceanography of the Red Sea. Oceanogr. mar. biol. A. Rev. 8:73–202.

Mortensen, Th., 1939. Two new Deep-sea Echinoderms from the Red Sea. Reports on the Preliminary Expedition for the Exploration of the Red Sea in the R.S.S. "Mabahith". Publs mar. biol. Stn Ghardaqa 1:37–46.

Moshkovitz, S. 1968. The Mollusca in the marine Pliocene and Pleistocene sediments of the southeastern Mediterranean basins (Cyprus-Israel). Thesis, Hebrew University of Jerusalem (English summary). 153 pp.

Moubayed, Z., 1986. Recherches sur la faunistique, l'écologie et la zoogeographie de trois réseaux hydrographiques du Liban: l'Assi, le Litani et le Beyrouth. Thesis. Université Paul Sabatier de Toulouse. 496 pp.

Moubayed, Z., & L. Botosaneanu, 1985. Recherches sur les Trichopteres du Liban et principalement des bassins superieurs de l'Oronte et du Litani (Insecta: Trichoptera). Bull. zool. Mus. Amsterdam 10(11):61–76.

Moubayed, Z. & M. Clergue Gazeau, 1985. Les Simuliidae (Diptera) des trois rivières: Orontes, Litani et Beyrouth. Annls Limnol. 21(1):83-88.

Moubayed, Z., N. Giani and E. Martinez-Ansemil, 1987. Distribution of Aquatic Oligochaeta and Aphanoneura in the Near East. In Krupp, F., W. Schneider & R. Kinzelbach (eds.), Proceedings of the Symposium on the Fauna and Zoogeography of the Middle East. Beihefte TAVO A 28:80–90.

Moubayed, Z. & H. Laville, 1983. Les Chironomides (Diptera) du Liban. I. Premier inventaire faunistique. Annls Limnol. 19(3);219–228.

Muerdter, D.R., J.P. Kennett & R.C. Thunell, 1984. Late Quaternary Sapropel Sediments in the Eastern Mediterranean Sea: Faunal Variations and Chronology. Quatern. Res. 21: 385–403.

Murina, V.V., 1971. Quantitative and qualitative features of the Red Sea macrobenthos. In Kisseleva, M.I. & A.A.Murina (eds.), The Benthos of the Red Sea Shelf, pp. 3–22. Naukova Dumka Kiev (in Ukrainian).

Myers, G.S., 1938. Freshwater Fishes and West Indian Zoogeography. Ann. Rep. Smithson. Instn :339–364.

Natterer, K., 1898. Expedition S.M. Schiff "Pola" in das Rothe Meer, nördliche Hälfte (October 1895 - Mai 1896) IX. Chemische Untersuchungen. Denkschr. Akad. Wiss., Wien 65:445–572.

Neev, D., N. Bakler & K.O. Emery, 1987. Mediterranean coasts of Israel and Sinai. Francis. N.Y., 130 pp.

Neev, D. & K.O. Emery, 1967. The Dead Sea. Bull. Geol. Surv. Israel 41:1–147.

Neev, D., L. Greenfield & J.K. Hall, 1985. Slice Tectonics in the Eastern Mediterranean. In Stanley, D.J. & F.-C. Wezel (eds.), Geological Evolution of the Mediterranean Basin, pp.249–269. Springer.

Neumann, A.C. & D.A. McGill, 1967. Circulation of the Red Sea in early summer. Deep Sea Res. 8:223–235.

Nieser, N. & Moubayed, Z., 1985. Les Heteroptères aquatique du Liban. I. Inventaire faunistique. Annls Limnol. 21(3):247–252.

Nilsson, T., 1983. The Pleistocene. Geology and Life in the Quaternary Ice Age. D. Reidel Publishing Comp. 529 pp.

Nir, Y., 1970. Les lacs quaternaires dans la région de Feiran (Sinai Central). Revue Géogr. phys. Géol. dyn. (2)12:335–346.

Nir, Y., 1973. Geological History of the Recent and Submarine Sediments of the Israel Mediterranean Shelf and Slope. Geol. Surv. Israel Mar. Geol. Div. Report. No. Mg. 10. 73. 135 pp.

Nir, Y., 1982. Asia, Middle East, Coastal Morphology: Israel and Sinai. In Schwartz, M.L. (ed.), The Encyclopedia of Beaches and Coastal Environments, pp. 86–98. Hutchinson Ross.

Ohlhorst, S. & G.E. Hutchinson, 1977. The waters of Merom: a study of Lake Huleh, 5. Temporal changes in the molluscan fauna. Arch. Hydrobiol. 80(1):1–19.

Ohlhorst, S., A. Schmida, M.M. Poulson & G. Evelyn Hutchinson, 1982. The waters of Merom: a study of Lake Huleh VIII, Non-siliceous plant remains, with appendices on some fossil animals. Arch. Hydrobiol. 94:441–459.

Oren, O.H., 1969. Oceanographic and biological influence of the Suez Canal, the Nile and the Aswan Dam on the Levant Basin. Rep. Progr. Oceanogr. 5:161–167.

Oren, O.H., 1970. The Suez Canal and the Aswan High Dam: Their Effect on the Mediterranean. Underwater Sci. Tech. J., December, 222–229.

Oren. O.H., 1983. Oceanography of the Mediterranean. In Karmon, Y., A. Shmuely & G. Horowitz (eds.), The Mediterranean Basin, pp. 87–100. Ministry of Defence Israel.

Ormond, R. & A. Edwards, 1987. Red Sea Fishes. In Edwards, A.J. & S.M. Head (eds.), Key Environments: Red Sea, pp. 251–287. Pergamon Press.

Ortal, R. & F.D. Por, 1978. Effect of hydrological changes on aquatic communities in the Lower Jordan River. Verh. int. Verein. theor. angew. Limnol. 20:1543–1551.

Ott, J.A., 1980. Growth and Production in *Posidonia oceanica* (L.) Delile. PSZNI Mar. Ecol. 1:47–64.

Pascar-Gluzman, C., 1981. A preliminary list of aquatic Oligochaeta from Israel: Naididae and Tubificidae. Israel J. Zool. 30:230–232.

Pascar-Gluzman, C. & Ch. Dimentan, 1984. Distribution patterns of Naididae and Tubificidae in Israel and the Sinai. Second Symp. on Aquatic Oligochaeta, Palanza Italy, September, 1982. Hydrobiologia 115:197–205.

Pasteur, R., V. Berdugo & B. Kimor, 1976. The abundance, composition and seasonal distribution of epizooplankton in coastal and offshore waters of the Eastern Mediterranean. Acta Adriat., Split. 18(4):55–80.

Paz, U., 1976. The rehabilitation of the Huleh Reserve. Israel Nature Reserves Authority. 80 pp.

Pearse, J.S., 1983. Signs of Stress on a Tropical Biota. In Abdel Latif, A.F., A. Refai Bayoumi & M-F. Thompson (eds.), Marine Science in the Red Sea. Bull. Inst. océanogr. Fish. Cairo, 9:148–159.

Peres, J.M., 1958. Ascidies recoltées sur les côtes Mediterranéennes d'Israel. Ascidies de la Baie de Haifa collectées par E. Gottlieb, Bull. Res. Counc. Israel 7B:143–164.

Peres, J.M., 1967. The Mediterranean Benthos. Oceanogr. mar. biol. A. Rev. 5:449 pp.

Peres, J.M., 1985. History of the Mediterranean Biota and the Colonization of the Depths. In Margaleff, R. (ed.), Key Environments, Western Mediterranean, pp. 198–232. Pergamon Press, Oxford.

Picard, L., 1963. The Quaternary of the northern Jordan Valley. Proc. Israel Acad. Sci. Hum. 1(4):1–34.

Picard, L., 1965. The geological evolution of the Quaternary in the Central Northern Jordan

Graben. Am. geol. Soc. So. Pap. 84:337–366.
Picard, L., 1970a. On Afro-Arabian Graben Tectonics. Geol. Rdsch. 59(2):2337–381.
Picard, L., 1970b. Further Reflections on Graben Tectonics in the Levant. "Graben Problems" International Upper Mantle Project, Scientific Report No. 27:249–267.
Plaziat, J.-Cl., 1982. Introduction a l'écologie des milieux de transition eau douce-eau salée pour l'identification des paleoenvironments correspondants. Critique de la notion de domaine margino-littoral. Mém. Soc. géol. Fr. 144:187–209.
Pollingher, U. 1978. The phytoplankton of Lake Kinneret. In Serruya, C. (ed.), Lake Kinneret, pp.229-242. Dr. W. Junk.
Pollingher, U., A. Ehrlich & S, Serruya, 1984. The Planktonic Diatoms of Lake Kinneret (Israel) During the Last 5000 Years - Their Contribution to the Algal Biomass. Proc. 8th Int. Symp. Living and Fossil Diatoms, Paris, 27–30 Aug. 1984 :459–470.
Por, F.D., 1962a. Un nouveau Thermosbaenacé, *Monodella relicta* n.sp. dans la dépression de la Mer Morte. Crustaceana 3(4):304–310.
Por, F.D., 1962b. *Typhlocirolana reichi* n.sp., un nouvel isopode cirolanide de la dépression de la Mer Morte. Crustaceana 4(4):247–252.
Por, F.D., 1964a. A study of the Levantine and Pontic Herpacticoida (Crustacea, Copepoda). Zool. Verh., Leiden No. 64:1–128.
Por, F.D., 1964b. The genus *Nitocra* Boeck (Copepoda Harpacticoida) in the Jordan rift valley. Israel J. Zool. 13(2):78–88.
Por, F.D., 1967. Level bottom Harpacticoida (Crustacea, Copepoda) from Elat (Red Sea), Part I. Israel J. Zool. 16(3):101–165.
Por, F.D., 1968a. The invertebrate zoobenthos of Lake Tiberias: I. Qualitative aspects. Israel J. Zool. 17:51–79.
Por, F.D., 1968b. The benthic Copepoda of Lake Tiberias and of some inflowing springs. Israel J. Zool. 17:31–50.
Por, F.D., 1968c. *Parabathynella calmani* n.sp. (Syncarida, Bathynellacea) from Israel. Crustaceana 14(2):151–154.
Por, F.D., 1968d. Solar Lake on the shores of the Red Sea. Nature, Lond. 218(5144):860–861.
Por, F.D., 1969a. Limnology of the heliothermal Solar Lake on the coast of Sinai (Gulf of Elat). Verh. int. Verein. theor. angew. Limnol. 17:1031–1034.
Por, F.D., 1969b. The Canuellidae (Copepoda, Harpacticoida) in the waters around the Sinai Peninsula and the problem of "Lessepsian" migration of this family. Israel J. Zool. 18(2–3):169–178.
Por, F.D., 1971a. One hundred years of Suez Canal - a century of Lessepsian Migration. Syst. Zool. 20(2):138–159.
Por, F.D., 1971b. The zoobenthos of the Sirbonian lagoons. Rapp. P.-v. Réun. Commn int. Explor. scient. Mer Méditerr. 20(3):247–249.
Por, F.D., 1972. Hydrobiological notes on the high-salinity waters of the Sinai Peninsula. Mar. Biol. 14(2):111–119.
Por, F.D., 1975a. Pleistocene pulsation and preadaptation of biotas in mediterranean seas: consequences for Lessepsian migration. Syst. Zool. 24:72–78.
Por, F.D., 1975b. An Outline of the Zoogeography of the Levant. Zool. Scripta 4:5–20.
Por, F.D., 1975c. The Coleoptera dominated fauna of the hypersaline Solar Lake (Gulf of Elat, Red Sea). In: Proc. 10th European Symposium on Marine Biology (Ostend, Belgium, Sept, 17–23, 1975), 2:563–573.
Por, F.D., 1978. Lessepsian Migration - the influx of Red Sea Biota into the Mediterranean by Way of the Suez Canal. Ecological Studies Vol. 23, Springer-Verlag, Berlin-Heidelberg-New York, 47 Figures, 10 Plates, 2 Maps, 228 pp.
Por, F.D., 1979. The Copepoda of Di Zahav Pool (Gulf of Elat, Red Sea). Crustaceana 37(1):13–30 + 4 plates.
Por, F.D., 1981. A Classification of the Hypersaline Waters based on Trophic Criteria. Mar. Ecol. (Naples) 1(2):121–131.
Por, F.D., 1982. A new species of *Bryocyclops* (Copepoda: Cyclopoida) and of *Parastenocaris* (Copepoda: Harpacticoida) from a cave in Israel and some comments on

the origin of the cavernicolous copepods. Israel J. Zool. 30(1):35–46.
Por, F.D., 1983a. The Lessepsian Biogeographic province of the Eastern Mediterranean. Rapp. Comm. Int. Mer Medit., Journées Études Syst. et Biogéographie Medit., Cagliari, 1980. pp. 81–84.
Por, F.D., 1983b. The Ecology of the Mediterranean coast of Israel and Sinai. In Schwartz, M. (ed.), Encyclopedia of Beaches and Coastal Environments, pp. 82–86. Hutchinson Ross.
Por, F.D., 1983c. The Freshwater Canthocamptidae (Copepoda: Harpacticoida) of Israel and Sinai. Israel J. Zool. 32(2–3):113–134.
Por, F.D., 1984a. The ecosystem of the mangal: General Considerations. In Por, F.D. & I. Dor (eds.), Hydrobiology of the Mangal - The Ecosystem of the Mangrove Forests. Developments in Hydrobiology. Vol. 20, pp. 1–14. Dr W. Junk Publishers, The Hague.
Por, F.D., 1984b. An Outline of the distribution patterns of the freshwater Copepoda of Israel and surroundings. Hydrobiologia 113:151–154.
Por, F.D., 1985. Anchialine pools - Comparative Hydrobiology. In Friedman, G.M. & W.E. Krumbein (eds.), Hypersaline Ecosystems: The Gavish Sabkha. Ecological Studies . Vol. 53, pp. 136–144. Springer.
Por, F.D., 1986. Crustacean biogeography of the Late Middle Miocene Middle Eastern landbridge. In Gore, R.H. & K.L. Heck (eds.), Crustacean Biogeography, pp. 69–84. Balkema Rotterdam/Boston.
Por, F.D., 1987. The Levantine Landbridge, Historical and Present Patterns. In Krupp, F., W. Schneider & R. Kinzelbach (eds.), Proceedings of the Symposium on the Fauna and Zoogeography of the Middle East. Beihefte TAVO A Nr. 28:23–28.
Por, F.D. & A. Ben-Tuvia, 1981. The Bardawil Lagoon (Sirbonian Lagoon) of North Sinai. A Summing up. Rapp. P.-v. Réun. Commn int. Explor. scient. Mer Méditerr. 27(4):101–107.
Por, F.D., H.J. Bromley, Ch. Dimentman, G.N. Herbst & R. Ortal., 1986. River Dan, headwater of the Jordan, an aquatic oasis of the Middle East. Hydrobiologia 134(2):121–140.
Por , F.D. & Ch. Dimentman, 1985. Continuity of Messinian Biota in the Mediterranean Basin. In Stanley, D.J. & F-C. Wezel (eds.), Geological Evolution of the Mediterranean Basin, pp. 545–557. Springer.
Por, F.D. & I. Dor, 1975. Ecology of the metahaline pool of Di Zahav, Gulf of Elat, with notes on the Siphonocladacea and the typology of near-shore marine pools. Mar. Biol. 29:37–44.
Por, F.D., I. Dor & A. Amir, 1977. The mangal of Sinai: Limits of an ecosystem. In Kinne, O. & H.-P. Bulnheim (eds.), International Helgoland Symposium "Ecosystem Research" (Helgoland, Sept. 1976). Vol.30(1–4), pp. 295–314. Helgoländer wiss. Meeresunters.
Por, F.D. & G. Eitan, 1970. The invertebrate zoobenthos of Lake Tiberias. II. Quantitative data (level bottoms). Israel J. Zool. 19(2):125–134.
Por, F.D. & Ruth Lerner-Seggev, 1966. Preliminary data about the benthic fauna of the Gulf of Elat (Aqaba), Red Sea. Israel J. Zool. 15(2):38–50.
Por, F.D. & D. Masry, 1968. Survival of a nematode and an oligochaete species in the anaerobic benthal of Lake Tiberias. Oikos 19:388–391.
Por, F.D. and R. Ortal, 1986. River Jordan - the survival and future of a very special river. Envir. Conserv. 12(3):264–268.
Por, F.D., H. Steinitz, W. Aron & Ilana Ferber, 1972. The Biota of the Red Sea and the Eastern Mediterranean (1967–1972) - a survey of the marine life of Israel and surroundings. "Contributions to the Knowledge of Suez Canal Migration". Israel J. Zool. 21(3–4):459–523.
Por, F.D. & M. Tsurnamal, 1973. Ecology of the Ras Muhammad Crack in Sinai. Nature, Lond. 241(5384):43–44.
Price, A.R.G., 1982. Arabian Gulf echinoderms in high salinity waters and the occurrence of dwarfism. J. nat. Hist. 16:519–527.
Price, A.R.G., P.A.H. Medley, R.J. McDowall, A.R. Dawson-Shepherd, P.J. Hogarth & R.F.G. Ormond, 1987. Aspects of mangal ecology along the Red Sea coast of Saudi

Arabia. J. nat. Hist. 21(2):449–464.
Purser, B.H., 1985. Coastal Evaporite Systems. In Friedman, G.M. & W.E. Krumbein (eds.), Hypersaline Ecosystems. The Gavish Sabkha, pp. 72–102. Springer.
Racek, A.A., 1974. The waters of Merom: a study of lake Huleh. IV. Spicular remains of freshwater sponges (Porifera). Arch. Hydrobiol. 74:137–158.
Raffi, S., 1986. The significance of marine boreal molluscs in the Early Pleistocene faunas of the Mediterranean area. Palaeogeogr. Palaeoclimat. Palaeoecol. 52:267–289.
Rampal, J., 1981. Biogéographie Mediterranéenne d'après l'étude des Thecosomes (Mollusques pelagiques). J. étud. Syst. évol. CIESM Cagliari, p.45.
Reiss, Z. & L. Hottinger, 1984. The Gulf of Aqaba. Ecological Micropaleontology. Springer. 345 p.
Reiss, Z., B. Luz, A. Almogi-Labin, E. Halicz, A. Winter & J. Erez, 1980. Paleoceanography of the Gulf of Aqaba (Elat), Red Sea. Quatern. Res. 14:294–308.
Remane, A. & E. Schultz, 1964. Die Strandzone des Roten Meeres und ihre Tierwelt. Kieler Meeresforsch. 20 (Sonderheft):5–17.
Riedl, R., 1966. Faunistische Studien am Roten Meer im Winter 1961/62. III. Die Aufsammlungen in Suez und Al-Ghardaqa, nebst einigen Bemerkungen über Gnathostomulida. Zool. Jb. 93:139–157.
Robba, E., 1987. The final occlusion of Tethys: Its bearing on Mediterranean benthic molluscs. In: K.G. McKenzie (ed.). 'Shallow Tethys 2", Balkema, Rotterdam, pp. 405–426.
Roberts, N., O. Erol, T. de Meester & H.F. Uerpman, 1979. Radiocarbon chronology of the late Pleistocene Konya Lake, Turkey. Nature, Lond. 281:662–664.
Rognon, P., 1979. Mecanismes climatiques actuels et paleoclimats au Sahara. Paleoecology of Africa and surrounding islands. Balkema, Rotterdam vol.11:1–12.
Rognon, P. 1980. Pluvial and arid phases in the Sahara: the role of non-climatic factors. Palaeoecology of Africa and surrounding islands. Balkema, Rotterdam vol.12:45–62.
Rosenan, D., 1951. Long range forecasts of rainfall. Bull. Res. Counc. Israel 1–2:33–35.
Rosenberg-Herman, Y., 1965. Études des sédiments quaternaires de la Mer Rouge. Thesis, Univ. Paris A(1123).
Rosenfeld, A., A. Segev & E. Halbersberg, 1981. Ostracode species and paleosalinities of the Pliocene Bira and Gesher formation (Northwestern Jordan Valley). Israel J. Earth Sci. 30:113–119.
Ross, D.A. & J. Schlee, 1977. Shallow structure and geological development of the southern Red Sea. Mineral Resourc. Bull. Jiddah 22.
Ross, J., J. Romero, E. Ballesteros & J.M. Gili, 1985. Diving in Blue Water. The Benthos. In Margaleff, R. (ed.), Key Environments. Western Mediterranean, pp. 233–295. Pergamon Press.
Rossignol-Strick, M., W. Nesteroff, P. Olive & C. Vergaud-Grazzini, 1982. After the deluge: Mediterranean stagnation and sapropel formation. Nature, Lond. 295:105–110.
Roth, G., 1987. Data on the Distribution and Faunal History of the Genus. *Theodoxus* in the Middle East (Gastropoda: Neritidea). In Krupp, F., W. Schneider & R. Kinzelbach (eds.), Proceedings of the Symposium on the Fauna and Zoogeography of the Middle East. Beihefte TAVO A 28:73–79.
Ruckert, F., 1985. Egel aus der Levante-Landern. Senckenberg. biol. 66(1–3):135–152.
Ryan, W.B.F., 1972. Stratigraphy of Late Quaternary Sediments in the Eastern Mediterranean. In Stanley, D.J. (ed.), The Mediterranean, a Natural Sedimentation Laboratory, pp. 149–169.
Saad, M.A., 1974. Calcareous deposits of the brackish water lakes in Egypt. Hydrobiologia 44:381–387.
Safriel, U.N., 1975. The Role of Vermetid Gastropods in the Formation of the Mediterranean and Atlantic Reefs. Oecologia (Berl.) 20:85–101.
Safriel, U.N. & U. Ritte, 1986. Suez Canal migration and Mediterranean colonization -- their relative importance in Lessepsian migration. Rapp. P.-v. Réun. Commn int. Explor. scient. Mer Méditerr. 29(5):259–263.
Said, R., 1981. The Geological Evolution of the River Nile. Springer. 151 pp.

Samaan, A.A. & A.A. Aleem, 1972. The Ecology of zooplankton in Lake Mariut. Bull. Inst. océanogr. Fish. Alexandria 2:375–397.

Sanlaville, P., 1982. Asia, Middle East, Coastal Morphology: Syria, Lebanon, Red Sea, Gulf of Oman and Persian Gulf. In Schwartz, M. (ed.), The Encyclopedia of Beaches and Coastal Environments, pp.98–102. Hutchinson Ross.

Sara, M., 1985. Ecological factors and their biogeographic consequences in the Mediterranean ecosystems. In Moraitou-Apostolopoulou, M. & V. Kiortsis (eds.,), Mediterranean Marine Ecosystems, pp. 1-17. Plenum Press.

Sauvage, H.E., 1882. Catalogue des poissons recueillis par M. E. Chantre pendant son voyage en Syrie, Haute Mesopotamie, Kurdistan et Caucase. Bull. Soc. philomath. Paris 7(4):163-168.

Scates, M.D., 1968. Notes on the Hydrobiology of Azraq Oasis, Jordan. Hydrobiologia 31:73–89.

Scheer, G., 1971. Coral reefs and coral genera in the Red Sea and Indian Ocean. Symp. zool. Soc. Lond. 28:329–367.

Scheer, G., 1984. The Distribution of Reef Coral in the Indian Ocean with a Historical Review of its Investigation. In Angel, M.V. (ed.), Marine Science of the Indian Ocean and Adjacent Waters. Proceedings of the Mabahiss John Murray Symposium, Egypt, 3–6 September 1983, pp. 885–900. Pergamon Press.

Scheer, G. & C.S.G. Pillai, 1983. Report on the stony corals of the Red Sea. Zoologica 133:1–198.

Schick, A.P., 1958. Tiran: the Straits, the Island and its Terraces. Israel Explor. J. 8(2–3):120–130; 189–196.

Schidlowsky, M.U., M. Matzigkeit, W.G. Mool & W.E. Krumbein, 1985. Carbon Isotope Geochemistry and C14 Ages of Microbial Mats from the Gavish Sabkha and the Solar Lake. In Friedman, G.M. & W.E. Krumbein (eds.), Hypersaline Ecosystems. The Gavish Sabkha, pp. 381–401. Springer.

Schlesinger, Y. & Y. Loya, 1985. Coral Community Reproductive Patterns: Red Sea Versus the Great Barrier Reef. Science, N.Y. 228:1333–1335.

Schmida, A., 1975. In Paz, U. (ed.), Nature Preservation in Israel, Researches and Surveys. Report 1, pp. 240–251. Israel Nature Reserves Authority.

Schminke, H.K., 1976. Systematische Untersuchungen an Grundwasserkrebsen. Eine Bestandsaufnahme (mit der Beschreibung zweier neuer Gattungen) der Familie Parabathynellidae (Bathynellacea). Int. J. Speleol. 8:195–216.

Schneider, W., 1987. The Genus *Pseudagrion* Selys 1876 in the Middle East. A Zoogeographic Outline (Insecta: Odonata: Coenagrionidae). In, W. Krupp, W. Schneider and R. Kinzelbach eds.: Proceedings of the Symposium on the Fauna and Zoogeography of the Middle East. Beihefte TAVO A 28, 114–123.

Schulman, N., 1959. Geology of the central Jordan Valley. Bull. Res. Counc. Israel 8G. 63–90.

Schumacher, H. & H. Mergner, 1985. Quantitative Analyse von Korallengemeinschaften des Sanganeb-Attols (mittleres Rotes Meer). II. Vergleich mit einem Riffareal bei Aqaba (nördliches Rotes Meer) am Nordrande des indopazifischen Riffgürtels. Helgoländer wiss. Meeresunters. 39:410–440.

Schumacher, H. and H. Zibrowius, 1985. What is Hermatypic? A Redefinition of ecological groups in corals and other organisms. Coral Reefs 4:1–9.

Schütt H., 1973. Die Mollusken eines jungpleistozänen Seeprofils im Becken von Damaskus. In: K. Kaiser, E.K. Kempf, Arl. Leroi-Gourhan and H. Schütt: Quartarstratigraphische Untersuchungen aus dem Damaskus-Becken und seiner Umgebung. Z. Geomorph. N.F. 17:3,319–335.

Schütt, H., 1983. Die Molluskenfauna der Süsswasser im Einzugsgebiet des Orontes unter Berücksichtigung benachbarter Flussysteme. Arch. Molluskenk. 113(1982) (1–6):17–91; 225–228.

Scott, R.W. & F.M. Govean, 1985. Early depositional history of a Rift Basin. Palaeogeogr. Palaeoclimat. Palaeoecol. 52:143–158.

Sengor, A.M.C., 1979. Mid-Mesozoic closure of Permo-Triassic Tethys and its implications.

Nature, Lond. 279:590–593.
Sengor, A.M.C., 1985. The story of Tethys: how many wives did Okeanos have? Episodes 8(1):3–12.
Serruya, C., 1978. Water Chemistry. In Serruya, C. (ed.), Lake Kinneret, pp. 185–204. Dr. W. Junk Publ.
Serruya, C. & U. Pollingher, 1983. Lakes of the Warm Belt. Cambridge Univ. Press. 569 pp.
Servant, M. & S. Servant-Vildary, 1980. L'environment quaternaire du bassin du Tchad. In Williams, M.A. & H. Faure (eds.), The Sahara and the Nile, pp. 133–162. Balkema, Rotterdam.
Shackleton, N.J., 1975. The stratigraphic record of deep-sea cores and its implications for the assessment of glacials, interglacials, stadials and interstadials in Mid Pleistocene. In Butzer, K.W. & G.L. Isaac (eds.), After the Australopithecines. 1–24.
Shackleton, N.J. & N.D. Opdyke, 1976. Oxygen isotope and paleomagnetic stratigraphy of Equatorial Pacific Core V 28–239, late Pliocene to latest Pleistocene. In Cline, R.M. & J.D. Hays (eds.), Investigations of Late Quaternary Palaeoceanography and Palaeoclimatology. Geol. Surv. Am. Mem. 145:449–464.
Shaw, H.F. & G. Evans, 1984. The nature, distribution and origin of sapropelic sediments of the Cilician Basin, northeastern Mediterranean. Mar. Geol. 61.1:1–12.
Sneh, A., 1974. Quaternary lakes in the northern Arava. Abstract. In Gill, D. (ed.), 1972–1973 Seminar of the Geological Survey of Israel.
Sneh, A. & G.M. Friedman, 1985. Hypersaline Sea-marginal Flats of the Gulfs of Elat and Suez. In Friedman, G.M. & W.E. Krumbein (eds.), Hypersaline Ecosystems. The Gavish Sabkha, pp. 103–135. Springer.
Sneh, A. & T. Weissbrod, 1973. Nile Delta: The defunct Pelusiac branch identified. Science, N.Y. 180:59–61.
Sonnenfeld, P., 1985. Models od Upper Miocene Evaporite Genesis in the Mediterranean Region. In Stanley, D.J. & F.-C. Wezel (eds.), Geological Evolution of the Mediterranean Basin, pp. 323–346. Springer.
Sonnenfeld P. & I. Finetti, 1985. Messinian Evaporites in the Mediterranean: A Model of Continuous Inflow and Outflow. In Stanley, P.J. & F.-C. Wezel (eds.), Geological Evolution of the Mediterranean Basin, pp. 347–354. Springer.
Sorbini, L., 1988. Biogeography and climatology of Pliocene and Messinian fossil fish of Eastern-Central Italy. Bull. Mis. civ. St. nat. Verona 14:1–85.
Sorbini, L. & R. Tirapelle-Rancan, 1980. Messinian fossil fish of the Mediterranean. Palaeogeogr. Palaeoclimat. Palaeoecol. 29:143–154.
Sournia, A., 1977. Notes on primary productivity of coastal waters in the Gulf of Elat (Red Sea). Int. Rev. ges. Hydrobiol. 62:813–819.
Spanier, E. & M. Goren, 1988. An Indo-Pacific trunkfish *Tetrosomus gibbosus* (Linnaeus): first record of the family Ostracionidae in the Mediterranean. J. Fish. Biol. 32:797–798.
Stanley, D.J., 1985. Mud Redepositional Processes as a Major Influence on Mediterranean Margin-Basin Sedimentation. In Stanley, D.J. & C.-F. Wezel (eds.), Geological Evolution of the Mediterranean Basin, pp. 337–410. Springer.
Steinhorn, I. & G. Assaf, 1980. The Physical Structure of the Dead Sea Water Column –1975–1977. In Nissenbaum, A. (ed.), Hypersaline Brines and Evaporitic Environments, pp. 145–153. Elsevier.
Steinitz, H., 1954. The distribution and evolution of the freshwater fishes of Palestine. Hidrobiologi Istanbul B1(4):225–275.
Steinitz, H., 1955. Occurrence of *Discoglossus nigriventer* in Israel. Bull. Res. Counc. Israel 5B:191–192.
Steinitz, H. & A. Ben-Tuvia, 1960. The Cichlid fishes of the genus *Tristramella* Trewavas. Ann. Mag. nat. Hist. 3:161–175.
Stephen, A.C., 1958. The Sipunculids of the Bay of Haifa and neighbourhood. Bull. Res. Counc. Israel 7B:129–136.
Steuer, A., 1942. Ricerche Idrobiologiche alle foci del Nilo. Memmorie Ist. ital. Idrobiol. 20:85–106.
Stiller, M., A. Ehrlich, U. Pollingher, U. Baruch & A. Kaufman. 1983–1984. The Late

Holocene sediments of Lake Kinneret (Israel) - Multidisciplinary study of a five meter core. GSI Current Research Geological Institute Jerusalem. pp. 83–88.

Stock, J.H., 1976. A new genus and two new species of the Crustacean order Thermosbaenacea from West Indies. Biejdr. Dierk. 46:47–70.

Stoffers P. & D.A. Ross, 1974. Sedimentary History of the Red Sea. In Whitmarsh, R.B., O.R. Weser & D.A. Ross (eds.), Initial Reports of DSDP 23, pp. 849–865. U.S. Govt. Printing Press.

Sturany, R., 1899. Lamellibranchiaten des Rothen Meeres. Expeditionen S.M. Schiff "Pola" in das Rote Meer (1895/96 und 1897/98). Zoöl. Ergebn. 14:1–41.

Sturany, R., 1903. Gastropoden des Rothen Meeres. Expeditionen S.M. Schiff "Pola" in das Rote Meer (1895/95 und 1897/98). Zoöl. Ergebn. 23:1–75.

Tchernov, E., 1971a. *Pseudamnicola solitaria* n.sp. a new prosobranch gastropod from the Dead Sea area. Israel J. Zool. 20:209–221.

Tchernov, E. 1971b. Freswater Molluscs of the Sinai Peninsula. Israel J. Zool. 20: 209–221.

Tchernov, E. 1973. On the Pleistocene Molluscs of the Jordan Valley. Israel Acad. Sciences and Humanities Jerusalem. 36 pp.

Tchernov, E. 1975a. The Molluscs of the Sea of Galilee. Malacologia 15(1):147–184.

Tchernov, E., 1975b. The Early Pleistocene Molluscs of 'Erq el Ahmar. Israel Acad. Sciences and Humanities Jerusalem . 36 pp.

Tchernov, E. 1987. The Age of the 'Ubeidiya Formation, an Early Pleistocene Hominid Site in the Jordan Valley, Israel. Israel Earth Sci. 36:3–36.

Tchernov, E., 1988. The biogeographical history of the Southern Levant. In Yom-Tov, Y. & E. Tchernov (eds.), The Zoogeography of Israel, pp. 159–250. Dr. W. Junk Publ.

Thiede, J., 1980. The Late Quaternary marine paleoenvironments between Europe and Africa. Palaeoecology of Africa 12:213–225.

Thiel, H., 1979. First quantitative data on Red Sea deep benthos. Mar. Ecol. Prog. Ser. 1:347–350.

Thiel, H., 1987. Benthos of the Deep Red Sea. In Edwards, A.J. & S.M. Head (eds.). Key Environments: Red Sea, pp. 112–127. Pergamon Press.

Thiel, H. & H. Weikert, 1984. Biological Oceanography of the Red Sea Oceanic System. In Angel, M.V. (ed.), Marine Science of the North-West Indian Ocean and Adjacent Waters. Proceedings of the Mabahiss John Murray International Symposium, Egypt, 3–6 September, 1983, pp. 829–831. Pergamon Press.

Thorson, G., 1957. Bottom Communities (Sublittoral and Shallow Shelf). Mem. geol. Soc. Am. 67:461–534.

Thorson, G., 1971. Animal migrations through the Suez Canal in the past, recent years and the future (a preliminary report). Vie Milieu, Supp. 22:841–846.

Thunell, R.C., 1979. Eastern Mediterranean Sea during the Last Glacial Maximum; an 18,000-Years B.P. Reconstruction. Quatern. Res. 11:353–372.

Tom, M., 1976. The Benthic Fauna Association of Haifa Bay. M.SC. Thesis, Tel Aviv University (in Hebrew).

Tortonese, E., 1951. I caratteri biologici del Mediterraneo orientale e i problemi relativi. Archo zool. ital., Suppl. 7:205–251.

Tortonese, E., 1985. Distribution and Ecology of Endemic Elements in the Mediterranean Fauna (Fishes and Echinoderms). In Moraitou-Apostolopoulou, M. & V. Kiortsis (eds.), Mediterranean Marine Ecosystems, pp. 57–83. Plenum Press.

Tsurnamal, M., 1968. Studies on the Porifera of the Mediterranean Littoral of Israel. Ph.D. dissertation. Hebrew University Jerusalem (Hebrew with English summary).

Tsurnamal, M. & F.D. Por, 1968. The subterranean fauna associated with the blind palaemonid prawn *Typhlocaris galilea* Calman. Int. J. Speleol. 3 (3–4):219–223.

Türkay, M., 1986. Crustacea Decapoda Reptantia der Tiefsee des Roten Meeres. Senckenberg. mar. 18(3/6):123–185.

Tzur, Y. & U.N. Safriel, 1978. Vermetid Platforms as Indicators of Coastal Movements. Israel J. Earth Sci. 27:124–127.

Vacelet, J., 1981. J. étud. Syst. Biogeogr. Méditerr. Cagliari, 29–30.

Valentine, J.W., 1967. The influence of climatic fluctuations on species diversity within the

Tethyan provincial system. In Adams, C.G. & D.V. Ager (eds.), Aspects of Tethyan Biogeography, pp. 153–166. The Systematic Assoc. London.
Van der Velde, I., 1983. Revision of African species of the genus *Mesocyclops* Sars, 1914 (Copepoda:Cyclopidae). Hydrobiologia 109:3–66.
Van Zeist, W. & H. Woldring, 1980. Holocene Vegetation and Climate of Northwestern Syria. Palaeohistoria 22:111–125.
Vanney, J.-R. & M. Gennesseaux, 1985. Mediterranean Seafloor Features: Overview and Assessment. In Stanley, J.P. & F.-C. Wezel (eds.), Geological Evolution of the Mediterranean Basin, pp. 3–32. Springer.
Vaumas de, E., 1957a. Sur l'évolution structurale et morphologique de la dépression du Ghab et du Bas-Oronte (Syrie). C.R. Acad. Sci. France 244, 24:2496–2498.
Vaumas de, E., 1957b. Sur la formation du résseau hydrographique de l'Oronte. Congr. Assoc. Francaise Avancement Sci. Paris.
Vergnaud-Garzini, C., 1985. Isotope Record: Stratigraphic and Paleoclimatic Implications. In Stanley, J.D. & F.-C. Wezel (eds.), Geological Evolution of the Mediterranean basins, pp. 413–452. Springer.
Vermeij, G. J., 1978. Biogeography and Adaptation. Harvard Univ. Press.
Vilwock, W., 1987. Further Contributions on Natural Hybrids between two Valid Species of *Aphanius, Aphanius dispar dispar* (Ruppell) and *Aphanius fasciatus* (Valenciennes), (Pisces:Cyprinodontidae) from the Bardawil-Lagoon, North Sinai, and al-Qanatir, West of the Suez Canal, Egypt. In Krupp, F., W. Schneider & R. Kinzelbach (eds.), Proceedings of the Symposium on the Fauna and Zoogeography of the Middle East. Beihefte TAVO A No. 28:238–244.
Volcani, E.B., 1936. Life in the Dead Sea. Nature, Lond. 138:467.
Volcani, E.B., 1940. Studies on the microflora of the Dead Sea. Ph.D. Thesis, Hebrew University of Jerusalem (Hebrew with English abstract).
Vries, C. de & G.W. Berendsen, 1954. Measurements of age by the Carbon-14 technique. Nature, Lond. 174, 4442:1131-1141.
Wainwright, S.A., 1965. Reef Communities visited by the Israel South Red Sea Expedition, 1962. Bull. Sea Fish. Res. Stn Israel 38:40–53.
Waisel, Y., 1967. A contribution to the knowledge of the phanerogamous vegetation of Lake Tiberias. Bull. Sea Fish. Res. Stn Israel 44:3–16.
Walker, D.I., 1987. Benthic Algae. In Edwards, A.J. & S.M. Head (eds.), Key Environments: Red Sea, pp. 152–168. Pergamon Press.
Weikert, H., 1982. The Vertical Distribution of Zooplankton in Relation to Habitat Zones in the Area of the Atlantis II Deep, Central Red Sea. Mar. Ecol. Prog. Ser. 9:129–143.
Weikert, H., 1987. Plankton and Pelagic Environment. In Edwards, A.J. & S.M. Head (eds.), Key Environments: Red Sea, pp. 90–111. Pergamon Press.
Weulersse, J.,1940. L'Oronte, Étude de Fleuve. Institut Francais. Damascus. 88pp.
Wewalka, G., 1986. Zoogeography and Ecology of the Dytiscidae Fauna of the Levant. Entomologica Basiliensia 11:273–288.
Williams, D.F., R.C. Thunell & J.P. Kennett, 1978. Periodic freshwater flooding and stagnation of the eastern Mediterranean Sea during the Late Quaternary. Science, N.Y. 201:252–254.
Williams. M.A.J. & D.A. Adamson, 1980. Late Quaternary depositional history of the Blue and White Nile rivers in Central Sudan. In Williams, A.J. & H. Faure (eds.), The Sahara and the Nile, pp. 207–224. Balkema, Rotterdam.
Williams, M.A.J. & F.M. Williams, 1980. Evolution of the Nile Basin. In Williams, M.A.J. & H. Faure (eds.), The Sahara and the Nile, pp. 207–224. Balkema, Rotterdam.
Winter, A., 1982. Paleoenvironmental interpretation of Quaternary coccolith assemblages from the Gulf of Aqaba (Elat), Red Sea. Revta esp. Micropaleont. 14:291–314.
Wishner, K.F., 1980. The Biomass of the Deep-Sea Benthopelagic Plankton. Deep Sea Res. 27:203–216.
Wolfart, R., 1967. Geologie von Syrien und dem Libanon. Beiträge zur regionalen Geologie der Erde, ed. H.J. Martin, vol. 6. Borntrager Berlin. 326 pp.
Wright, R., 1979. Benthic foraminiferal repopulation of the Mediterranean after the

Messinian (late Miocene) event. Palaeogeogr. Palaeoclimat. Palaeoecol. 29:189–214.

Yaron, Z., 1964. Notes on the ecology and the entomostracan fauna of temporary rainpools in Israel. Hydrobiologia 24:489–513.

Yom-Tov, Y. & H. Mendelssohn, 1988. Changes in the distribution and abundance of vertebrates in Israel during the 20th century. In Yom-Tov, Y. & E. Tchernov (eds.), The Zoogeography of Israel, pp. 515–548. Dr. Junk Publ.

Zahran, M.A., 1977. Africa A. Wet Formations of the African Red Sea Coast. In Chapman, V.J. (ed.) Ecosystems of the World, pp. 215–232. Elsevier Amsterdam.

Zalcman, Dorit & F.D. Por, 1975. The food web of Solar Lake (Sinai coast, Gulf of Elat). Rapp. P.-v. Réun. Commn int. Explor. scient. Mer Méditerr. 23(3):133–134.

Zeuner, F.E., 1959. The Pleistocene Period. Hutchinson London. Zibrowius, H., 1977. Inventaire des Scleractiniaires de la Mediterranée. Rapp. P.-v. Réun. Commn int. Explor. scient. Mer Méditerr. 24/4:183–184.

Zibrowius, H., 1980. Les Scleractinaires de la Mediterranée et de l'Atlantique nord-occidental. Mém. Inst. océanogr. Monaco 11:1–227.

Zibrowius, H. & G. Bitar, 1981. Serpulidae (Annelida, Polychaeta) indo-pacifiques établis dans la region de Beyrouth, Liban. Rapp. P.-v. Réun. Commn int. Explor. scient. Mer Méditerr. 27/2:159–160.

Zibrowius H. & A.A. Ramos, 1983. *Oculina patagonica*. Scleractinaire exotique en Mediterranée. Nouvelles observations dans le sud-est de l'Espagne. Rapp. P.-v. Réun. Commn int. Explor. scient. Mer Méditerr. 28.

Zohary, M. & G. Orshansky, 1947. The Vegetation of the Huleh Plain. Palest. J. Bot. Jerusalem Ser. 4:90–104.

Zwick, P., 1972. *Protonemura zernyi* Aubert (Insecta: Plecoptera), an addition to the fauna of Israel. Israel J. Zool. 21:49–51.

Zwick, P., 1984. *Marthamea beraudi* (Navas) and its European congeners (Plecoptera:Perlidae). Annls Limnol. 20 (1/2):129–139.

Geographical and Subject Index

Adiabatic heating 93
Adriatic Sea 41, 72, 78
Aegean Sea 17, 25, 41, 49, 61, 78
Afar Triangle 31
Africa (North Africa, East Africa, etc.) 13, 14, 19, 25, 30, 31, 137
African biota 121, 126, 127, 128, 137–141 ff
Afrin 18
Afrotopical biota 137
Agricultural drainage
    Nile 63, 101
    Jordan 121
Aharne swamps 134
Akko (Acres) 45, 47
Aleppo (Halab) 35, 131
Alexandria 63
Amanus mountains 136
Amatzia 39
Amictic lake 123
Amiq Golu (see Lake Amiq)
Ammiq swamps (Lebanon) 144
Anaerobic conditions 4, 26, 27 ff, 73, 93, 123, 124, 134
Anatolia (Anatolian Mountains) X, 27, 32, 45, 141, 145, 159
Anchialine environments 8, 51, 91–97 ff
Antarctica 7, 12
Antakiya 110, 135, 136
Antilebanon mountains 110, 144, 148
Antilessepsian migration 73, 75
Aqaba 12, 66, 87
Aquaculture 163
Aquatic "steeple-chase" 14, 18, 20, 37, 38, 48, 129, 131–132 ff, 134
Arabia 5, 6, 19, 39, 139, 140
Arabian Sea 29, 72, 86
Aridity (see also Desertification) 30, 39
Asiatic biota (included West Asiatic)
Ashmura formation 36
Askalon 143
Aswan 14
Aswan High Dam 31, 45, 53, 62–64 ff
At Tur 66, 73, 79

Baalbek 133

Bab el Hawa swamp lake 158
Bab el Mandab Straits 18, 78, 84, 89
Badenian 3
Bahr el Ateibe 37
Bahr el Hijjan 37
Balearic Sea 9
Ballah lagoon 102
Basaltic flows 18, 34–36 ff, 148 ff
Bay of Akkar 47
Bay of Haifa 27, 50
Beaufort castle 144
Beer Sheva 4
    depression 14
Bekaa Valley 18, 35, 38, 142–145 ff, 147
Belayim lagoon (Ghor Blayim, Al Belayim) 11, 29, 97
Benot Yaaqov bridge 149
Bermuda 54
Bet Shean valley 35, 113, 118, 120
    lake 36, 121
Bet Zaida (B'tekha) valley 152
Beyrouth 31, 52
Bikini attol 89
Bira formation 14, 15, 114
Birket Mamilla 159
Birket Ram 109
Birket Yamouneh (see Lake Yamouneh)
Bitter Lakes 28, 29, 60, 72, 98–100 ff, 103
Black Sea 3, 25, 26, 41, 48, 59, 107
BOD (Biological Oxygen Demand) 106
Bodenhemier Line 155
Boreal biota 20, 21
Bosphorus, Straits of 8, 27
Brackish environments 6, 7, 27, 91, 103, 113, 121, 138, 153, 156
    biota 14
Brunhes Normal Magnetic Polarity 22, 36
Burdigalian 3, 4, 137

Caesarea 143
Calabrian 20, 21
Cambrian 1
Cambridge Expedition to the Suez Canal 102–3
Canary Islands 54
Cape Guardafui 81

Caribbean Sea 4, 6
Carmel ridge 47
Caspian Sea 3, 127
Caucasus Mountains 141
Caucasian biota 145
Cenozoic 83
Cheizar 134
Circum-mediterranean biota 139, 140
Circumtropical biota 59
Cisjordanian rivers 153, 158
Cis-Rift mountains 37, 142, 153
CLIMAP 24
Climatic Optimum 37
Clysmic Gulf 5, 7, 9, 12
Coastal Plain (of Israel) 30, 137, 138, 159
Coastal streams 38, 103–107 ff, 153
    biota 119, 126, 133, 137, 145, 146, 163
    history 38
    hydrology 144, 153
    pollution 105, 106, 153
Cold-stenothermic biota 141, 142, 150, 163
"Coral community" 79, 89, 99
Coral reefs 11, 12, 27, 50, 55, 66, 79, 85–90 ff
Coralligenous community 50
Coriolis force 31
Crab community 53
Crenal 146, 148
Crete, Island of X, 11, 25, 26, 41
Cretaceous 1, 2, 4, 83, 126
Crocodile River (see also Nahal Tanninim) 105
Cyprus, Island of 11, 42, 43, 47, 59
Cyrenaica 41

Dahab (Di Zahav) 94
Damietta branch 31
Damietta Cone 45
Damascus Basin 32, 34, 37, 38, 129, 130, 131
Daphne 109
Dead Sea 6, 12, 32, 109, 112, 114–118 ff, 119, 120, 139, 156
    biota 114
    history 14, 18, 20
    hydrography 113, 116, 153
    lake levels 36, 115, 119, 160
Deglaciation (see also Termination) 8, 30
Deep Sea (Levant Basin and Red Sea) 66
    biota 17, 27, 52, 81, 83–85 ff
    history 13, 27
    hydrography 17, 27, 45, 48
Desert pools 6, 157
    streams 6, 155–156

Desertification (see also Aridity) 7, 30, 31, 32, 116, 123, 127, 137
Diatomite 113
"Discovery" cruise 66
Dispersal passive (see Passive Dispersal)
Dolina's 144

East Siberiaan biota 142
Easter Island 89
Eastern Ghor Channel 121
Eddies 43, 44
Egypt 30, 46
Eilat (Elat) 66, 87, 88, 89, 91
El Arish (see Wadi El Arish)
El Azraq (Qa el Azraq)
El Bab, River 35
El Ghardaqa 66, 79
El Gisr 24
El Kura lagoon 92, 97
El Qardud 97
En Avedat 118, 156
En Fashkha 113, 119, 163
En Fidje 110
En Fuliya 118
En Gedi 113
En Mzarib 110
En Qudeirat, 155
En Safi 119
En Yabrud 4
En Yahav 118
En Zarqa 110, 133, 144
Enot Samar 118
Endemic biota
    Mediterranean 10, 11, 13, 49 ff
    Red Sea 27, 80–85 ff
    Continental waters 111, 119–120, 129, 131, 146, 151, 152
Endorheic basins 18, 19, 34, 37, 38
Ephemerous waters (Temporary waters) 112, 156–159 ff
Er Rastan sill (see Rastan)
Eremic biota 139, 140, 141, 151
Erq el Ahmar lake 36
Ethiopian biota 112, 114, 128, 137–141 ff, 147, 151, 154, 155, 159
    Lakes 30, 31.155
Euphrates (and Tigris basin) XI, 3, 4, 37, 38, 109, 111, 127, 128, 129 ff, 131, 134, 136, 139
Euro-siberian biota 141, 142
Euryhaline biota 5, 9, 11, 14, 15, 73, 111, 113, 119, 120, 125–126 ff, 127, 137, 142, 147, 151, 156
Eustatic sea levels 21–23, 27, 29, 30, 46, 73, 89, 138
Eutrophic conditions 77

biota 29
Eutrophication 126
Evaporites 9, 12
Euxinic conditions 27

Faraun Island 97
Fars evaporites 3
Fernando de Noronha Islands 54
Fertile Crescent 127
Fish ponds 149, 163
Fisheries 61 ff, 63–63, 101, 112, 134, 136, 163
Flandrian 24, 30, 96, 98
"Freshwater Tethys" 127, 137

Gadot lake 148
Galilee 150, 153
Gavish Sabkha 96 ff
Gaza 31
Gebel Druz 131
Gebel Hauran 110
Gebel Katarina (Santa Katharina mountain) 155
Gemmulae 125
Ghab valley 18, 35, 38, 39, 109
"Ghor" 97, 98
Gibraltar 1, 8, 17, 55, 90
Glacials, 21–26, 73, 82, 90, 141
Golan Heights 4, 146, 150, 153, 156, 157, 161, 162 ff
Gondwanian biota 5, 6, 125, 137
Greek Islands X
Guardafui, Cape 81
Guemsah lagoon 97
Gulf of Aden 27, 29, 67, 77, 80
Gulf of Aqaba (Gulf of Eilat) 66–73 ff
  biota 51, 75, 78–79 ff, 83–84, 85
  history 12, 18, 20, 29, 30
  hydrography 65 ff
Gulf of Genua 55
Gulf of Guinea 22
Gulf of Iskenderun (Gulf of Alexandretta) 38, 45, 47
Gulf of Oman 81
Gulf of Suez 11, 55, 96, 98, 99
  biota 74, 78, 79 ff, 89
  history 5, 12, 28, 29
  hydrography 65

Haifa 45, 47
Halmyric environments 91–107, 118
  biota 119
Hama 133, 134, 136
Hamei Zohar 113, 119
Hasbani (see River Hasbani)
Hasbaya 142

Hatay 3
Hauran, Gebel 34, 110
Hawaii 93
Headwater capture (see also Aquatic steeple-chase) 37, 38, 131, 142
Hellenic ridge 8
Hermatypic corals 56, 87–91 ff
Hermel 133
Hermon, Mount (see also Antilebanon) 110
Himalaya 6
Holocene 31
Homs 35, 38, 136
Hormuz, Straits of 127
Hot brines 29, 66, 68, 83
Hot springs 113 ff
Hula Nature Reserve, 149, 152
Hydraulic changes 105, 118, 121–122, 134, 142, 143, 159, 160–162 ff
Hyperarid conditions 30
Hypersaline environments 6–11, 27, 30, 68, 91, 94 109, 112, 119, 159
  biota 94, 96, 98, 99
Hyporheic environment 142, 146

India 30
Indian Ocean 8, 11, 12, 17, 18, 48, 65, 68, 74, 75, 80, 81, 83, 89, 90, 93
Indoasiatic biota 128
Indomalayan biota 127
Indopacific biota 4, 12, 18, 49, 55, 56, 60, 61, 73, 83, 85
Infralittoral biota 50, 53
Interglacials 21–29
Interstitial biota 79
Intertidal biota 61, 75
Introduced species 135, 147, 163 ff
Ionian Sea 9, 25, 26, 41, 61
Irrigation 116
Isthmus of Suez 5, 12, 17, 18, 24, 98 ff, 101
"Isthmian" biota 74 ff, 98

Jeddah 66, 68
Jerusalem 6, 143, 159
Jezireh 3, 4,
Jordan (see River Jordan)
Jordan Rift Valley (see also Rift Valley) 7, 14, 18, 19, 32, 35, 129, 131, 137, 138, 139, 156
Jordanian desert 129
Judea 156
Judean mountains 15
Jurassic 1, 3, 110

Kareem formation 5

Karkor sill 134, 135
Karst 144, 146
Kebara swamps 105
King Talal Dam 140
Kurkar, aeolianitic sandstone 46–47 ff, 54

"Lago Mare" 9
Lagoons 28, 29, 68, 74, 97, 100, 102
Lake Amiq (Amiq Golu, Lake of Antakiya or Antiochia)109, 135
   history 15, 35, 38 ff, 109, 131, 135, 136
   biota 136 ff
Lake Burullus 100, 101
Lakes, Central African 30, 31, 123
Lake Edku 100, 102, 103
Lake Homs (Lake Qatina) 109, 133, 134
Lake Hula (Lake Huleh) 32, 148–152 ff
   biota 136, 139, 151–152 ff
   drainage of 148, 149, 152
   history 34, 36 ff, 112, 148
   hydrography 149–151
Lake Kinneret(Lake Tiberias, Sea of Galilee) 6, 35, 109, 112, 118, 119, 121, 122 ff, 152, 161, 163
   biota 114, 125–126 ff, 135, 139, 140, 151, 158, 162, 163
   history 34, 36, 113, 122
   hydrography 122–125 ff, 149, 160
Lake Konya 32
Lake levels 30, 31, 32, 116, 122
Lake Lisan 35–37 ff, 112, 119, 121, 122, 148, 156
Lake Manzalah 100, 101
Lake Mariut 100, 101
Lake Mobutu 30
Lake Qarun (Lake Karun) 109
Lake Saif 39
Lake Tana 155
Lake Turkana 30
Lake Victoria 30
Lake Yamouneh (Birket Yamouneh) 109
   hydrology 144, 145
   biota 146
Lebanese shore 54
Lebanon 133, 137, 138, 140, 143–148, 152
Lebanon mountains 129, 143–148
Lessepsian migration 42, 48, 49, 53, 55–62 ff, 73
Lessepsian migrant biota 13, 51, 64 ff, 81, 102
Lessepsian province 41, 61–62 ff
Levant Basin 41–42 ff
   benthos 17, 48 ff
   biota 17, 49, 50
   coasts 45–48 ff
   history 3, 24–25
   hydrography 24–25, 42–45 ff
   plankton 48, 52–53 ff
Levantine Intermediate Water 42
Libya 25
Litani (see River Litani)
Lower Lacustrine Bed (Hula) 148

"Mabahiss" expedition 66
Madaba map 116
Maharda Dam 134
Main Peat Unit (Hula) 148
Malta, Island of 64
Mangal 75–77 ff
Mangrove 66, 75–77 ff, 81, 97
"Manihine" expedition 66
Mansoura 144
"Marsa" 97
Matapan Cape 41
   trench 41
Mediterranean-Dead Sea Canal, 117, 118
Mega-Chad 31
Meiobenthos 48, 51, 66, 78, 79, 94
MESEDA (Red Sea metalliferous bottoms project) 84
Mesopotamian biota 112, 126–129 ff, 141, 142, 146
Messinian 4, 7, 8, 9, 12, 51, 137
   biota 9 ff, 11
   relics 9
Metahaline environments 28, 29 ff, 65–68 ff, 82, 91, 94, 97 ff, 101
   biota 94, 96, 97, 101, 107
"Meteor" expedition 84
Metulla-Majayoun area 18
Midwater fishes 84
Milazzian 21, 22
Mindel 21, 22
Miocene 3, 4, 5, 6, 7, 9, 13, 23, 51, 118, 127, 129, 137
Monastirian 21, 22
Monomictic lake 123
Monsoons 32, 67, 78, 81
Monte Bolca slates 82
Monte Castellaro 11
Moss fauna 6

N-T water (Noit-Tiberias waters) 113 ff, 119
Nab'l Bardauni stream 145
Nabq 97
Nahal Alexander 105
Nahal Daliya 105
Nahal Dan (see River Dan)
Nahal Hadera 105
Nahal Hermon (see River Banias)
Nahal Iyon 148

Nahal Naaman 105
Nahal Oren 105
Nahal Qishon 105, 138
Nahal Soreq 159
Nahal Tanninim 105, 107
Nahal Yarqon 105, 153
Nahal Zin 156
Nahr Aouali 144, 145
Nahr Barada 110
Nahr Beyrout 144
Nahr Damur 144, 145
Nahr el Kebir North 38, 138
Nahr el Kebir South 38, 133, 138
Nahre el Kelb 147
Nahr Hasbani (see River Hazbani)
Nahr Marqiya 134
Nahr Qab Elias 145
Nahr Zarqa (Jordan) 110
Nannism, Levantine 49 ff
National Water Carrier of Israel 121, 160–163 ff
Negev 6, 32, 39, 119, 120, 139, 155, 157, 158, 162
"Negev Fauna" 137
Nehring Line 150, 153 ff, 155
Neogene 128
Nile (see also River Nile) 14 ff, 138, 140
  current (flow) 26, 30, 42, 47, 48, 51, 55, 62, 63, 64, 118
  sediments 45 ff, 50
Nilotic biota 138, 139, 152

Olduvai Normal Event 21
Oligocene 4, 5
Oligohaline biota 105, 120, 123, 124
Oligomictic lake 123
Oligotrophic conditions 30, 43, 73, 77, 78, 83
Oriental biota 127, 129
Orontes (see River Orontes)
Oxygen Minimum Layer 71–72 ff, 78

Palaearctic biota 112, 127–129, 140, 141–143 ff, 150, 151, 153–156 ff
Palaeoeremic biota 140
Paleoeuropean biota 128
Palaeotropic biota 127, 137
Paleogene 4
Paleotethys 2
Paleozoic 1, 2
Pan-Mediterranean biota 51
Palmyra oasis, ridge (Tadmur) 4, 129
Palynology 37, 39 ff
Panama Isthmus, Canal
Paratethys 3, 7, 8, 9, 10, 127

Passive dispersal 57, 127, 137, 139, 140, 155
Pelusiac branch 101
Perim, Straits of 27
Persian Gulf 11, 18, 29, 65, 68, 72, 80, 131
Petra 155
Phreatic environments 6, 141
Piacenzian(Plaisancian) 13, 14, 15, 20
Pikre Limne 98
Planktobenthos 51
Pleistocene 5, 8, 17, 17–30 ff, 51, 112, 137, 148, 151
  chronology 20–24 ff
  climates 17
  pulsations 23, 32, 67–68 ff, 109
Pliocene 4, 5, 6, 8, 12–15 ff, 17, 290, 81, 83, 112, 114, 118, 119, 121, 125, 125, 127, 129, 137
Pluvials and Interpluvials 30, 31, 151
"Pola" expedition 42, 52, 66, 83, 84
Polluted waters 66, 105, 106, 121, 134, 137, 145, 149, 159–160 ff, 161, 162
Ponor's 144
Ponto-Caspian IX, 11, 129, 141
Potamal 146
Port Sudan 66, 68, 89
Port Suez 89
Precambrian 1
Pre-lessepsian biota 18, 55
Primary freshwater biota 5, 111, 126–127 ff, 141
Prototethys 1

Qa el Azraq 4, 34, 110, 120–121 ff, 130, 131, 159, 163
Qa el Jafr 37
Qanat's 110, 143
Quaternary 17, 20, 38
Qumran 116

Rain pools 156–159 ff
Ras Atantur 76
Ras Gharib lagoon 96
Ras Kanisa lagoon 97
Ras Lahata lagoon 97
Ras Matarma lagoon 97
Ras Muhammad (sabkha and reef crack) 93, 94, 96
Ras Shukeir 79
Rastan 35, 133, 134
Red Sea IX, XI, 65 ff
  benthos 74–77, 79, 82–84
  fishes 80–82, 83, 84
  history 5, 9, 11, 12, 17, 27
  hydrography 65–66, 71

plankton 78
productivity 77
provinciality 67, 68, 73–80 ff, 86
tides 67, 87
Reefs (see Coral reefs)
Reef cracks 93, 94
Relics
   Gondwanian 5, 6
   Messinian 9, 10
   Miocenic 7, 51, 118
   Pliocenic 118
   Tethyan 8, 10
Reservoirs (storage lakes) 135, 160–163
Resting eggs 157
Rhithral 146, 150
Rhodes, Island of 17, 25, 41, 42, 49, 54, 59
Rift Valley (Dead Sea, Jordan, Syrian.etc.) IX, XI, 14, 18, 109, 112, 119, 120, 125, 132, 136, 47, 148
Rifting (see also Tectonics) 17–19, 34 ff, 109
Rhomb grabens 19, 109
Riss 21, 39
River Afrin 18, 38, 109, 131, 135, 136
River Banias (Nahr Banias, Nahal Hermon) 144, 148, 149, 151 ff
River Ceyhan 38, 131, 136
River Dan 144, 148, 149, 150
   hydrology 110, 150
   biota 150–151 ff, 163
River Hazbani (Nahr Hasbani, Nahal Senir) 38, 144, 148, 149
River Jordan 109, 112 ff, 114, 143, 148, 163
   biota 121, 122, q29–131 ff, 148, 163
   history 37, 112, 131
   hydrology 112, 121, 148, 149, 160, 161
   irrigation 121
   pollution 121
River Kara Su 18, 38, 109, 131, 136
River Kucuk Asi (Ak Su) 38, 135, 136
River Litani (Nahr Litani) 35, 133, 144–147 ff
   biota 129, 130, 138, 141
   history 35, 37, 38, 109
   hydrology 110
   pollution
River Nile (see also Nile flow, Nile sediments, Nile delta) X, 5, 17, 24, 30, 31 ff, 137, 139
River Orontes (Nahr Asi) 109, 133–137 ff, 143, 144, 145, 146
   biota 129–131 ff, 138, 139, 142
   history 15, 19, 35, 38 ff
   hydrology 110
   pollution 134

River Qwaik 35, 38
River Yarmoukh (Nahr Yarmoukh) 110, 121
Rock pools 157
Rocky bottom fauna (Mediterranean) 50, 51
Rosetta branch 31
Rosh Haniqra (Ras el Naqura) 50
Roudj 134
Rudeis formation 5

SRSJC (Sudanese Red Sea Joint Commission) 29
Sabkha's (see also Hypersaline environments) 9, 96, 97
Sahara 30, 32, 39, 137, 138, 139, 140
Sahel 30
Samaria 143, 156
Samra Lake 35
Sanganeb reef 87
Sapropel 25, 26 ff, 31, 48, 52, 90
Saudi Arabia 3
Sea grasses, 29, 49, 77 ff, 99
Sea of Marmara 41
Secundary freshwater biota 5, 111
Sedom formation 14
Senegalian biota 21, 23, 53 ff, 55
Serapeum 24
Sharm 97
Shubaq 155
Shurat el Arwashie mangal 75
Sicily 64
Sicilian 24
Sidra, Sea of (Gulf of Sidra) 41, 53
Sinai Peninsula XI, 19, 30, 46, 91, 92, 129, 139, 155, 156
   freshwaters of 32, 155–156 ff, 157, 158
Sindh 129
Sirbonic lagoon (see Bardawil lagoon)
Solar Lake 91, 94
   history 94
   hydrography 93, 95
   biota 94–96
Somali Coast 81
"Sonne" cruise 66
Springs
   artesian 110 ff, 131, 144, 150
   saline 6, 113, 118–121, 124, 139
Spring oases 32, 132, 138, 155, 156
Sponge bottoms 50
Statoblasts 158
Stranded fauna 51
Stratification
   Gulf of Aqaba 77, 78
   Dead Sea 116
   Lake Kinneret 123

Stream zonation 146 ff
Stromatolites 11
Submarine caves 48, 51, 94
Subterranean fauna 6, 7, 41, 118–119, 140, 142
Subtroglophilic biota 153
Sudanese Red Sea 87
Sudd swamps 31
Suez Canal X, XI, 4, 12, 55–62 ff, 98–100ff
   history 98–99
   hydrography 60, 63
   migration (see also Lessepsian migration) 13, 42, 53, 55 ff 59, 81
Surface Peat Unit (Hula) 148
Syria 137
Syrian Desert 129

Tabgha springs 118
Tanitic branch 101
Taurus Mountains 141
Tectonics (see also Rifting) 4, 5, 8, 17, 22, 30, 34, 35, 38, 39, 48, 51, 53, 55 ff, 59, 81, 142
Termination (see also Deglaciation) 22, 23, 24, 26, 29, 30, 31
Tethys Sea IX, 1–5, 6, 10
Tiran, Straits of 71, 84
Tortonian 7, 10
Towartit reef 87
Transgression 6, 7, 12, 29, 126
Transjordanian desert 120, 139
   rivers 153
Trans-Rift mountains 142, 153, 155
Travertines 39

Triassic 1
Tripoli formation 7
Trubi transgression 12
Turkey 61
Tyrrhenian 21, 22, 24
   biota 53 ff, 55

Ubeidiya Lake 36, 112
Upper Lacustrine Phase (Hula) 148

"Valdivia" cruise 66
Vardar Sea 3
Vermetide platforms 54 ff

Wadi Arbain 155
Wadi el Arish 38, 107, 138, 155
Wadi Feiran 155
Wadi Fidan 34
Wadi Kelt 153
Wadi Mujib 119
Wadi Shaag 155
Wadi Sirhan 4, 18, 34
Wadi Taal 155
Wadi Tumilat 101
Wurm 21, 30, 34, 37, 39

Yarda basalts 36
Yemen 139
Yezreel Valley 14, 18, 138

Zagros mountains 141
Zanclian 12, 13, 14, 15
Zerqa Main 113
Ziqlag formation 4
Zooxanthellate animals 29, 50, 56

Authors Index

Aartsen van, J.J. 57
Adamson, D.A. 30.31
Agur, Z. 60
Aharoni, I. 134
Aleem, A.A. 63, 99, 100, 102
Almaca, C.59
Almeida Prado-Por, M.S. 78, 105, 107
Almogi-Labin, A. 73
Alouf, N.J. 142, 144, 145
Amittai, A. 73
Andrea, H.G.85
Arad, A. 7, 14
Aron, W. 78, 84
Ayal, Y. 49

Bacescu, M., 27, 52
Baibars (Sultan), 146
Baird, W. 153, 156
Balss, H. 83
Banarescu, P. 37, 127, 128, 129, 153
Baranes, A. 85
Barash, Al. 57, 59, 61, 74, 81, 138
Barrois, Th. 156
Bartov, H. 59
Baruch, U. 39
Bebars, M.I. 64
Beckmann, W. 77
Begin, B.Z. 32, 35, 113
Ben Avraham, Z. 20, 65
Ben-David, Z. 105
Ben-Eliahu, N.M. 59, 61
Ben-Tuvia, A. 48, 58, 59, 61, 62, 101, 126, 152, 163
Benayahu, Y. 87
Bender, F. 4, 34, 113
Benson, R.H. 1, 12
Bentor, Y. 45
Berendsen, G.W. 35
Berdugo, V. 52
Berg, L.S. 127
Berggren, A.A. 30
Berman, T. 63
Bernoulli, D. 1, 2
Berthelmy, C. 145, 146
Beyth, M. 116
Bizon, G. 12

Boletzki S.V. 49
Bonatti, E. 68
Botosaneanu, L. 120, 140, 145, 146, 153, 155
Braithwaite, C.J.R. 89
Briggs, J.C. 80
Brinkhurst, R.O. 123
Bromley, H.J. 118, 119, 120, 140, 142, 146, 154, 155,
Brown, E.S. 155
Buchbinder, B. 3, 17
Bruins, H. 32
Butzer, K.W. 22, 30

Calvert, S.E. 26
Campbell, A.C. 81
Carey, W.S. 164
Carrozza, F. 57
Casanova, J.P. 52, 78
Cernosvitov, L. 142
Chalifa, Y. 4
Chappuis, P.A. 155
Chen, H. 140
Cherbonnier, G. 59
Chervinski, J. 138
Christiaens, J. 94
Clark, A.M. 81
Clerge-Gazeau, M. 142
Coad, B.W. 127
Cohen, J. 96
Cowgill, U. 32, 36
Crosskey, R.W. 142
Crossland, C. 66

Danin, Z. 57, 59, 61, 74, 81
Darom, D. 50, 54, 75, 76, 79, 85, 90, 125, 158
Degens, E.T. 29, 68
Delibrias, J. 31
Demoulin, G. 142
Deuser, W.G. 29
Dexter, D.M. 51
Dia, A. 140, 142, 144, 145, 146
Dimentman, Ch. 7, 9, 10, 11, 36, 118, 119, 125, 127, 139, 142, 148, 149, 150, 151, 152, 156, 157, 159

Dor, I. 92, 94, 96, 150
Dowidar, N.M. 63
Dubertret, L. 134, 136
Dumont, H.J. 32, 137, 138, 139, 152, 155

Eckstein, Y. 32
Edwards, A.J. 74, 82
Edwards, F.J. XI, 68
Ehrlich, A. 96, 113, 148, 150
Eitan, G. 123, 125
Ekman, S. 1, 4, 80, 81
El-Heyawi, M.L.E. 63
El-Maghraby, A.M. 63
El-Sabh, M.L. 64
Elsayed, A.I.W. 97
El-Sharkawy, S.M. 100
Elster, H.J. 102
El-Wakeel, S. 101
Emery, K.O. 14, 32, 35, 36, 45, 116
Eren, Y. 147
Esteban, M. 11
Evans, G. 27, 29,
Ezzat, A. 102

Fabricius, F.K. 13
Fairbridge, E.W. 32
Farrand, W.P. 32, 37, 39
Finetti, I. 8, 9
Fishelson, L. 87
Fleminger, A. 59
Forsskal, P. 66
Fouda, M.M. 74, 75
Francheteau, J. 17
Fredj, G. 13, 27, 42, 48, 49, 52,
Fricke, H. 85, 89
Friedman, A. 115
Friedman, G.M. 12, 96, 97
Fryer, G. 6
Fuchs, Th. 24
Furnestin, M.L. 52, 60

Gaillard, J.M. 49
Galil, B. 59, 61
Garfunkel, Z. 5, 19, 109
Gasith, A. 120, 140
Gasse, F. 31
Gat, J.R. 25, 32, 113
Gaudy, R. 52
Gavish, E. 91, 96
Gennesseaux, M. 8, 45
George, R.Y. 27, 52
Gerdes, G. 91, 95, 96
Gilat, E. 50, 51, 53, 61
Girdler, R.W. 5, 9, 12, 68, 109
Godeaux, J. 78
Golani, D. 59, 60, 61

Golandsky, B. 77
Goodyear, R.H. 78, 84
Goren, M. 59, 126
Gorgy, S. 45, 50
Gottlieb (see Gilat)
Govean, F.M. 5, 9
Grecchi, G. 18
Gruvel, A. 136
Guergues, S.K. 32, 60
Gwirtzman, G. 3, 4, 17

Haas, G. 4, 5, 138
Haeckel, E. 66
Halim, Y. 60, 63, 78
Hammerton, D. 63
Hartland-Rowe, R. 156
Head, S.M. 65, 66, 68, 80, 82, 86, 87
Hedgpeth, J.W. 91
Hellal, A.M. 74, 75
Hemprich, K. 66
Herbst, G.N. 7, 106, 108, 117, 118, 142
Herman, Y. 52
Herrnstadt, I. 150
Hey, R.W. 23
Heyn, C.C. 150
Holthuis, L.B. 49, 53, 91, 92
Hopkins, T.S. 42
Horowitz, A. 4, 14, 15, 18, 20, 22, 32, 36, 39, 112, 148
Hottinger, L. 29, 30, 70, 71, 72, 73, 77, 89
Hsu, K.J. 8, 11
Hubault, F. 6
Hulsemann, K. 59
Hutchinson, G.E. 36, 123

Illies, J. 146
Inbar, M. 149
Issar, A. 32

Jach, M.A. 120, 139, 142, 151, 159
James, D.B. 79
Jamieson, B.G.M. 123
Jones, D.A. 76, 91
Jones, R.F. 150
Josephus Flavius, 115

Kafri, U. 7, 14
Kaiser, K. 32, 37
Kaplan, I.R. 115
Karaman, G.S. 139
Karmon, Y. 148, 149
Kaufman, A. 35
Kempf, E.K. 37
Khalil, A.N. 63
Khouri, R.G. 160
Kiener, A. 91

Kimor, B. 52, 77
Kinsman, J.J. 11
Kinzelbach, R.K. 8, 19, 37, 38, 120, 121, 127, 128, 129, 130, 131, 133, 134, 136, 138, 139, 147
Kiortsis, V. 41
Kisseleva, M.I. 78
Klausewitz, W. 27, 80, 81, 82, 84, 85, 93
Klein, C. 115, 116
Klein, M. 159
Klinker, J.Z. 71, 77
Klunzinger, C.B. 66
Kocatas, A. 61
Koch, S. 6
Kronfeld, J. 39
Kosswig, C. 10
Koster, B. 130
Krumbein, W.E. 96
Krupp, F. 4, 5, 20. 37. 120. 126, 127, 129, 131, 134, 137, 138, 140, 142, 147
Kugler, J. 122, 140
Kuhlmann, D.H.H. 89
Kukla, G. 22, 23, 24
Kullenberg, B. 26

Laborel, J. 54
Lacombe, H. 17, 43
Lakkis, S. 52, 53, 59
Lasserre, G. 64
Lattin, G. de 138
Laubier, L. 13, 27, 42, 48, 50, 52
Laubscher, H. 1, 2
Laville, H. 140
Le Pichon, X. 17
Lepsi, I. 116
Lernau, H. 138
Lerner-Seggev, R. 51, 65, 79, 83, 123
Leroi-Gourhan, A. 37
Levanon-Spanier, I. 77
Leventer, H. 161, 162, 163
Levi, Cl. 49
Lesseps, F. de 164
Lewinsohn, Ch. 61, 80
Liebmann, E. 31
Linnavuori, R. 120, 140, 151, 155
Lipkin, Y. 77, 150
Liere, Van, W.J. , 35, 38
Livermore, R.A. 1, 3
Livnat, A. 39
Locard, A. 134
Loffler, H. 156
Lortet, L. 15, 136
Lotan, R. 118
Loya, Y. 87, 89
Luksch, J. 66
Luz, B. 73

Magaritz, M. 25, 31
Mancy, K.H. 63
Mangold, K. 49
Marenzeller, E. von 52
Margaleff, R. 41, 42
Margalit, Y. 120, 155
Martens, 139
Martinez-Ansemil, E. 142
Masry, D. (see Darom, D.)
Mastaller, M 81, 82, 83
Mazloum, S. 132
Mazor, E. 112, 113, 118, 119
McClure, H.A. 31
McGill, D.A. 67
McKenzie, J.A. 7, 8, 164
Mendelssohn, H. 138, 152, 159, 161
Menzies, R.J. 26, 27, 52
Mero, F. 113, 123
Mergner, H. 66, 73, 87, 88
Meulenkamp, J.E. 17
Mienis, H. 81, 105, 107
Molcho, M. 118
Mor, D. 18
Moraitou-Apostolopoulou, M. 52, 53, 59
Morcos, S.A. 64, 72
Mortensen, Th. 83
Moshkovitz, S. 21
Moubayed, Z. 134, 140, 142, 144, 145, 146, 147
Muerdter, D.R. 25
Munns, R.G. , 100
Murina, V.V. 78
Myers, G.S. 126

Natterer, K. 66
Neev, D. 1, 14, 30, 32, 34, 35, 36, 116
Neumann, A.C. 67
Nieser, N. 142
Nilsson, T. 22
Nir, Y. 44, 45

Oberhansli, 8
Ohlhorst, S. 32, 36, 151
Opdyke, N.D. 20, 22
Oren, O.H. 31, 41, 46, 62
Ormond, R. 73, 81
Orshanski, G. 147
Ortal, R. 119, 138, 159, 160, 161
Ott, J.A. 49

Pascar-Gluzman, C. 142
Pasteur, R. 52
Paz, U. 46
Pearse, J.S. 79
Peres, J.M. 49, 53
Picard, L. 18, 32, 36, 50, 148

Pillai, C.S.G. 87
Pizanty, S. 105
Plaziat, J-Cl. 91
Pollingher, U. 101, 126
Price, A.R.G. 68, 76
Purser, B.H. 96

Racek, A.A. 32, 36, 125
Raffi, S. 21
Ramos, A.A. , 55
Rampal, J. 53
Raz, G. 73
Reiss, F. 140
Reiss, Z. 4, 29, 30, 68, 71, 72, 73, 77, 89
Remane, A. 79
Riedl, R. 79
Ritte, U. 60
Robba, E. 4
Roberts, N. 32
Rognon, P. 39
Rosenan, D. 32, 39
Rosenberg-Herman, Y. (see also Herman)
Rosenfeld, A. 114
Ross, D.A. 9, 17, 29, 68
Ross, J. 49
Rossignol-Strick, M. 26
Roth, G. 37, 123, 127, 147
Rowe, F.W.E. 81
Ruckert, F. 142
Ruppell, E. 66
Ryan, W.B.F. 27

Saad, M.A. 102
Safriel, U.N. 54, 60
Said, R. 14, 30
Samaan, A.A. 102
Sanlaville, P. 47
Sara, M. 49
Scates, M.D 131, 159
Scheer, G. 79, 82, 86, 87
Schick, A.P. 89
Schidlowski, M.U. 96
Schlee, J. 9, 17
Schlesinger, Y. 87
Schmida, A. 150
Schminke, H.K. 6
Schneider, W. 20, 126, 139, 140, 152
Schulman, N. 14
Schultz E. 14
Schumacher, H. 55, 73, 87, 89, 120
Schutt, H. 5, 37, 131, 142, 148
Scott, R.W. 9
Seggev, R.(see Lerner-Seggev, R.)
Sengor, A.M.C. 1, 2, 164
Serruya, C. 101, 122, 122, 123
Servant, M. 31

Servant-Vildary, S. 31
Shackelton, N.J. 20, 22
Sharaf El Din, S.H. 100
Shaw, H.F. 27
Smith, A.G. 1, 3
Sneh, A. 39, 97, 101
Sonnenfeld, P. 3, 8, 9, 11, 17
Sorbini, L. 10, 11, 12
Sournia, A. 77
Spanier, E. 35, 59
Stanley, D. J. 13, 26
Steinitz, G. 18
Steinitz, H. 5, 152
Stephen, A.C. 49
Steuer, A. 102
Stiller, M. 126
Stock, J.H. 6
Stoffers, P. 29
Sturany, R. 66, 83
Suess, E. 1
Svoboda, A. 78

Tahori, A.S. 155
Tchernia, P. 17, 44
Tchernov, E. 13, 18, 36, 38, 112, 120, 129, 137, 155, 156
Thiede, J. 24. 25
Thiel, H. 78, 81, 84
Thorson, G. 53, 98
Thunell, R.C. 24, 25, 26
Tirapelle-Rancan, R. 10, 11, 12
Tom, M. 61
Tortonese, E. 10, 13, 49, 52
Tsoar, H. 32
Tsurnamal, M. 51, 91
Turkay, M. 83, 84
Tzur, Y. 54

Vacelet, J. 11
Valentine, J.W. 81
Van der Velde, I. 126
Van Zeist, W. 39
Vanney, J-R. 8, 45
Vaumas, E. de 35, 133
Vergnaud-Garzini, C. 13
Vermeij, G.J. 60
Vilwock, W. 120
Volcani, E.B. 115
Vollenweider, R.A. 102
Vries, C. de 35

Wainwright, S.A. 89
Waisel, Y. 125
Walker, D.I. 82
Weikert, H. 71, 72, 77, 78, 83
Weissbrod, T. 101

Weulersse, J. 108, 109, 132, 133, 134, 136
Wewalka, G. 136, 139, 151
Williams, D.F. 26
Williams, F.M. 14
Williams, M. A. J. 14, 30
Winter, A. 73
Wishner, K.F 83
Woldring, H. 39
Wolfart, R. 4, 15, 35
Wright, R. 18

Yaron, Z. 153
Yom-Tov, Y. 138, 159, 161

Zahran, M.A. 75
Zalcman, D. 83, 96
Zeuner, F.E. 21
Zibrowius, H. 27, 55
Zohary, M. 150
Zwick, P. 142

Taxonomic Index

abalone 79
Acabaria 100
Acanthobrama (=Mirogrex partim) 128, 131
Acanthobrama lissneri 138
Acanthobrama terraesanctae 163
Acanthobrama tricolor 131
Acanthopleura haddoni 79
Acartia clausi 102
Acartia fossae 59
Acartia latisetosa 102
Acartia negligens 52
Achaeus erythraeus 84
Actinopyga mauritiana 79
Afrocyclops gibsoni 121, 139, 151
Afronurus 140
Afronurus kugleri 151
Agaricidae (Scleractinia) 79
Albunea carabus 53
Alburnus 128
Alburnus selal 139
Aloidis gibba 51
Alpheidae (Decapoda) 51
Amphibia 152, 158
Amphipoda (Crustacea) 6, 102, 111, 119, 121, 142, 145, 150
Amphisorus hemprichi 29
Amphiura chiajei 51
Ancylus fluviatilis 147
Anguilla 15
Angulus valtonis 102
Anhinga melanogaster chantrei 136
Anostraca 127, 156, 159
Antecaridina lauensis 93
Antedon mediterranea 51
Aphanius 11, 12, 114, 120
Aphanius dispar 10
Aphanius d.dispar 120
Aphanius d.richardsoni 120
Aphanius fasciatus 120
Aphanius mento 120, 126, 151
Aphanius sirhani 120, 131
Appendicularia 52, 53
Appendicularia sicula 53
Arbacia lixula 50
Arctica islandica 20

Arctodiaptomus 140
Arctodiaptomus similis 125, 156, 158
Arcularia gibbosula (=Nassa) 74
Aristichthys nobilis 163
Arius thalassinus 59
arrow worms, see Chaetognatha
Artemia 96, 159
Ascidiacea 49, 53, 95
Astatotilapia flavijosephi 126, 138
Asterina burtoni 60, 68
Astronesthes martensii 84
Astropecten polyacanthus 68
Atrioplanaria aquaebellae 154
Athanas amazona 53
Atherina boyeri 12
Attheyella crassa 154, 155
Attheyella naphtalica 155
Atyaephyra desmarestii 19, 121, 131, 163
Atyaephyra d.orientalis 120, 130
augers 81
Avicennia marina 75, 76

Balanus improvisus 102
Barbus 127, 128
Barbus canis 131
Barbus chantrei 131
Barbus grypus 131
Barbus longiceps 131
Barbus luteus 131
Barbus pectoralis 131
barncles 102
Basilichthys bonairensis 163
Belostoma cordofanum 151
Belostomatidae (Hemiptera) 140, 151
Bithynia phialensis 147
Bivalvia 16, 56, 83, 127, 133, 142, 147, 150
Blenniidae (Pisces) 82, 126
Blepharicera fasciata 145
Blepharoceridae (Diptera) 145
bluegreen algae 96, 116
Bogidiella hebraea 6, 119
Borelis melo 4
Bosmina coregoni maritima 102
Bosmina longirostris 126, 139, 151, 163
Brachidontes variabilis 56, 60, 102

Branchinecta ferox 156
Branchinella spinosa 159
Branchiostoma lanceolatum 51
Branchipus schaefferi 156, 158
brine flies 118
brine shrimps 96
brittle stars 50
Bruguiera gymnorhiza 75
Bryocamptus minutus 154
Bryocyclops absalomi 6
Bryophyta 150
Bryozoa 50, 105, 158
Bryssopsis lyrifera 51
Buccinum undatum 20
Bufo viridis 159
Bulinus truncatus 131, 138

Cabillus anchialinae 93
Caenis macrura 155
Calanipeda aquae-dulcis 107
Calanoidea (Copepoda) 51, 52, 53, 78, 94, 125, 139
Calanus helgolandicus 52, 59
calcareous red algae 50
Calliasmata pholidota 93
Callinectes sapidus 105
Callionymus pussilus 11
Calocalanus pavo 53
Candona sp. 114
Cantharus fumosus 83
Canthocamptidae (Harpacticoidea) 125, 142, 150, 154, 155
Canthocamptus microstaphylinus 154, 158
Canthydrus diophthalmus 139
Canuella perplexa 102, 105, 107
Capoeta 128
Capoeta damascina 129, 134, 147
Carcharinus brevipinna 59
Cardita aculeata 51
Caridina 127
Caridina fossarum syriaca 129
carp 163
Cassidaria echinophora 51
Cassiopeia andromeda 56, 94
Caulerpa scalpelliformis 50
Centropages kroyeri 52
Centropages typicus 52
Centropages violaceus 52
Cephalopoda 49
Cerastoderma glaucum 74, 102
Ceratophyllum demersum 102
Ceriodaphnia reticulata 151
Ceriodaphnia rigaudi 126, 139
Cerithidea cingulata 77
Cerithium caeruleum 60
Cerithium scabrium 59, 60, 102

Cervinia 83
Chaetodontinae (Pisces) 81, 82
Chaetodon jaykari 84
Chateognatha 52
Chara vulgaris 163
Charophyta 125
Charybdis longicollis 61
Chirocephalus appendicularis 158
Chirocephalus bairdi 156, 158
Chirocephalus neumanni 156, 158
Chironomidae (Diptera) 125, 140, 145, 146, 151
chitons 50
Chlamys islandicus 20
Chlorophyta 50, 126
Chondrostoma 127, 128,
Chondrostoma regium 137
Chydoridae (Cladocera) 156
Cichlidae (Pisces) 126, 127, 128, 138, 151, 163
Ciliata 116
Cirolanidae (Isopoda) 51
Cladocera 102, 111, 126, 139, 156, 163
Cladocora caespitosa 27, 50, 55
Clarias gariepinus 126, 128, 134, 138, 163
Clavatula nifat 53
Clementia papyracea 56
Cleopatra bulimoides 138
Cletocamptus deitersi 121
Cloeon dipterum 118, 155
cobia 59
Cobitinae (Cyprinidae) 125, 141, 142, 146, 150, 151, 153, 155
Cobitis sp. 146
Coccolithophorida 29, 73
Coelenterata 94
Coleoptera 96, 97, 111, 138, 139, 141, 142, 146, 150, 151, 159
Conchostraca 127, 156
Congeria 9
Conidae (Gastropoda) 81
Conus testudinarius 22
Corallium rubrum 49
Corbicula fluminalis 127, 129, 133, 147, 163
cordgrass 102
Cordylophora caspia 102
Corophium 102
Corophium orientale 107
Cortispongilla barroisi 125
cowries 81
crabs 19, 83, 1155
Crassostrea 97
Crenilabrus 53
Crenobia 146
Cresseis acicula 53, 73, 78

Crinoidea 80, 82
Crocodylus niloticus 137, 138
Crocothemis sanguinolenta 140
Cryptocladopelma virescens 125
Cryptocyclops linjanticus 139
Ctenopharyngodon idella 163
Culex mimeticus 155
Culicidae (Diptera) 151, 155
Cuspidaria 83
Cyanophyta 126
Cyathura carinata 105
Cyclopoidea (Copepoda) 6, 120, 121, 125, 126, 156, 163
Cyclosalpidae (Thaliacea) 52
Cymathium 49
Cypraea 81
Cypraea lurida 53
Cypreidae (Gastropoda) 81
Cyprideis littoralis 10
Cyprideis torosa 114, 120, 125
Cyprinidae (Pisces) 6, 126, 127, 141, 146, 163
Cyprinodontidae (Pisces) 126
Cyprinus carpio 135, 163
Cystoseira myrica 77
Cyzicus sp. 156

damselflies 139, 152
Daphnia sp. 102, 140, 156
Daphnia atkinsoni 156
Daphnia lumholtzi 126, 139
Darwinula africana 139
Darwinula cf. stevensoni 125
Decapoda (Crustacea) 6, 49, 51, 53, 61, 80, 83, 84, 111, 126, 141
Dendrocoelum 146
Dendropoma petraeum 154
Dentalium dentale 51
Diadema setosum 79
Diamesinae (Chironomidae) 146
Diamysis bahirensis 10, 102, 107
Diamysis bahirensis hebraica 105
Diamysis bahirensis sirbonica 102
Diaphanosoma excisum 102
Diaptomidae (Calanoidea) 127, 141, 156, 158, 159
Diatomacea 113, 126, 150
Dicentrarchus 97
Dicentrarchus labrax 102
Dicentrarchus punctatus 74
Dichotrix eylathensis 82
Dina sp. 155
Dina shtschegolewi 154
Dinoflagellata 126, 163
Diodora yaroni 94
Diodora yaroni isaaci 94

Diogenes pugilator 54
Diptera 97, 111, 120
Discoglossus nigriventer 152
Doliolum nationale 78
dragonflies 139, 151, 152
Dreissena 127
Dugesia 146
Dugesia biblica 151, 154, 155
Dugesia b.biblica 120
Dugesia salina 120
Dunaliella parva 116
Dytiscidae (Coleoptera) 120, 139

Echinodermata 29, 49, 79, 81
Echinocardium cordatum 51
Echinogammarus foxi 107, 119
Echinogammarus veneris 119, 121, 125
Echinoidea 81, 83, 87
Echinometra mathaei 79, 97
Echinothrix calamaris 79
Ecteinascidia 94
edible blue crabs 105
eels 105, 136
Eichhornia crassipes 138
Elaphoidella 155
Elasipodidae (Holothuroidea) 52
Elmidae (Coleoptera) 146, 150
Engraulis enchrassicolis 74
Eocyzicus sp. 156, 158
Ephemeroptera 111, 112, 140, 142, 145, 146, 150, 151, 155
Ephydridae (Diptera) 118
Epinephelus sp. 12, 53
Eretes sticticus 139
Erpobdellidae (Hirudinea) 142
Eucalanus crassus 78
Eucyclops serrulatus 154
Eucypris clavata 159
Eudorylaimus andrassyi 123
Euphausiacea 52
Eurycletodes 83
Euterpina acutifrons 102

fairy shrimps 157
Ferissia aff.wautieri 127
fidler crabs 76
fish (see Pisces)
Fissurella nubecula 53
flatworms 141, 151, 155
Foraminifera 5, 18, 21, 26, 29
Fromia ghardaqana 79

Gambusia affinis 153
Gammarus 102, 142
Gammarus syriacus 154
Garra 127, 128

Garra rufa 126, 138, 151
Garra tibanica ghorensis 5
Garra variabilis 137
Gastropoda 22, 49, 57, 77, 81, 83, 96, 114, 127, 138, 142, 147, 163
Gastrotricha 79, 111, 141
Gephyrocapsa oceanica 73
Gerris 150
Globigerinidae (Foraminifera) 52
goatfish 61
Gobiidae (Pisces) 10, 59, 82, 93
Gobius 11
Gorgonaria 27, 50, 100
grey mullet 105
Gyrinidae (Coleoptera) 142

Hadziidae (Amphipoda) 51
Halicyclops 120
Haliotis pustulatus 79
Halobacterium 116
Halodule uninervis 77
Halophila ovalis 77, 79
Halophila stipulacea 18, 29, 49, 77
Halophiloscia couchii 125
Haloptilus longicornis 78
Harpacticoidea (Copepoda) 5, 51, 83, 96, 192, 119, 121, 125, 142, 150, 151, 158
Heliporus steindachneri 84
Helobdella stagnalis 155
Hemidiaptomus gurneyi canaanita 156
Hemigrammocapoeta 128
Hemigrammocapoeta culiciphaga 131, 134
Hemigrammocapoeta festai 147
Hemigrammocapoeta nana 129, 131
Hemiptera 97, 111, 129, 138, 141, 150, 151, 155
Hemiramphus cf. far 12
Herpetocypris chevreuxi 159
Herpetocypris salinarum 120
Heterocentrotus mamillatus 79
Heteroptera 142
Hippocampus brevirostris 74
Hippopotamus amphibius 137, 138
Hirudinea 126, 142
Holothurioidea 26, 59, 81
Homarus gammarus 49
Hyalinea baltica 21
Hydraenidae (Coleoptera) 120
Hydrobia 114
Hydrobiidae (Gastropoda) 10, 120, 125
Hydroidea 102
Hydroides 59
Hydromedusae 52
Hydrophilidae (Coleoptera)120, 139
Hydroptila simulans 120
Hyla arborea 159

Hypalocletodes 83
Hyperiidea (Amphipoda) 52
Hypophthalmichthys molitrix 163

Ilyocypris sp. 114
Iranocichla hormuzensis 127
Irpa (=Kolga) ludwigi 52
Ischnura senegalensis 139
Isidiella elongata 50
Isopoda (Crustacea) 6, 51, 111, 121, 155, 159

Jago omanensis 84

keyhole limpets 93
Kinorhyncha 79
Kolga(=Irpa) ludwigi 52

Labidocera detruncata 53
Labidocera madurae 59
Lanistes sp. 138
Lates 137
Lates macrophthalmus 138
Leda pella 51
leeches 140, 141, 155
Leguminaia saulcyi 131
Leguminaia wheatleyi 131, 133
Lepidurus apus apus 156, 157, 158
Lepidurus a. lubocki 156, 158
Leptocaris brevicornis 125
Leptoseris 89
Leuciscus 128, 131
Leuciscus cephalus 134
Leuciscus spurius 134
Limacina inflata 53
Limnatis nilotica 140
Limnogeton fieberi 151
Lithophyllum tortuosum 54
Liza aurata 74
loaches 128, 150, 153
lobsters 49
Lophelia pertusa 27
Lotilia gracillosa 82
Loxoconcha galilea 125
Lutjanus argentimaculatus 59
Lymnea auricularia 163
Lymneidae (Gastropoda) 142
Lynceus cf. brachyurus 156

Macoma calcarea 20
Macrobrachium rosenbergi 153
Macropipus pussilus 54
Macrothrix 156
Mactra olorina 56, 102
Madrepora oculata 27
Maja goltziana 53
Malleus regula 56

Mastacembellus simach 128, 135
Matuta banksi 59
Medusae 94
Melanoides tuberculata 120, 127, 129, 131, 156, 163
Melanopsis 9
Melanopsis nodosa 129
Melanopsis praemorsa costata 131, 163
Melanopsis p. buccinoidea 147, 154
Mesocletodes 83
Mesocyclops sp.102
Mesocyclops leuckarti 126
Mesocyclops ogunnus 126, 139, 163
Mesophyllax aspersus 153, 155
Metacrangonyx 6
Metadiaptomus chevreuxi 139, 158
Metapenaeopsis judaensis 59
Microcyclops 156
Microcyclops minutus 156
Micronecta 140
Micronecta scutellaris 151
Micropanope rufopunctata 53
Micropterna 153
Middendorfia caprearum 50
Mitra fusca 53
Modiolus auriculatus 60
Moina 140, 156
Moina dubia 102
Moina ? micrurum 102
Molgulidae (Ascidiacea) 49
Mollusca 5, 18, 20, 21, 29, 56, 57, 59, 60, 61, 81, 97, 102, 111, 121, 125, 126, 131, 136, 137, 138, 140, 141, 147, 156
Monodella relicta 6, 119
Moraria 155
mud skippers 77
Mugilidae (Pisces) 97, 102
Mullus barbatus 61
Mullus surmuletus 61
Munida japonica 83
Muraenesox cinereus 59
Muraenidae (Pisces) 81
Murex tribulus 57, 83
Mustellus mosis 84
Mya truncata 20
Myllopharygodon piceus 163
Myriophyllum spicatum 125
Mysidacea (Crustacea) 10, 102
Mytilus 57

Naididae (Oligochaeta) 150
Nais bretscheri 142
Najas marina 125, 163
Nannopus palustris 125
Naseus nuchalis 82
Nassarius albescens 83

Natica flammulata 51
Nemacheilus 146
Nemacheilus abyssinicus 155
Nemacheilus insignis 129
Nematoda 111, 123, 125
Nematopagurus helleri 84
Nematopagurus lewinsohni 84
Nematopagurus longicornis 84
Nemertea 79
Neogloboquadrina dutretei 26
Neolovenula alluaudi 139, 156, 157, 158
Nereis diversicolor 102, 105
Niphargus 142, 145
Niphargus nadarini 150
Nitocra balnearia 119
Nitocra incerta 121, 151
Notostraca 127, 156, 158
Nun jordanicus(=Orthrias jordanicus) 153

Oculina patagonica 55
Ocypode cursor 53
Odonata 111, 139, 140, 141, 155
Oligochaeta 111, 126, 141, 142, 145, 150
Oligoneuriopsis 140
Oniscoidea (Isopoda) 125
Onychocamptus mohammed 107, 121, 125
Ophiocoma pica 79
Ophirion 83
Ophiura texturata 50
Ophiuroidea 26, 52
Opisthobranchia 57
Oratosquilla massawensis 61
Orchestia cavimana 119
Orchestia platensis 119
Orthetrum ransonetti 139
Orthetrum taeniolatum 139
Orthrias (see Nun)
Ostracoda 10, 26, 111, 114, 120, 125, 139, 156, 159
Oxyurichthys papuensis 59

Pachygrapsus transversus 53
Paguristes calvus 83
Paguristes incomitatus 83
Palaemon ellegans 102, 107, 121, 122
Palemonidae (Decapoda) 51
Palaeostoma mirabile 83
Palinurus elephas 49
Paphia textile 56
Parabathynella 6
Paracalanus aculeatus 53
Paracalanus parvus 52
Paragomphus sinaiticus 139
Paramonacanthus oblongus 81
Parapandalus adensameri 84
Parastenocaris 6

Pavona 79
pearl oysters 18
Pectinura vestita 52
peixe rei 163
Pelecypoda (see Bivalvia)
Peloscolex kurenkovi 142
Pelobates syriacus 159
Pempheridae (Pisces) 59
Pempheris vanicolensis 59
Pennatula rubra 51
Peracarida (Crustacea) 126
Periclimenes pholeter 93
Pericosmus akabanus 83
Peridinium cinctum 126, 163
Periophthalmus 76
Petrobiona massiliana 11
Phanerogamia 150
Phoxinellus 128, 148
Phoxinellus kervillei 146
Phoxinellus libani 146
Phoxinellus syriacus 146
Phoxinellus zeregi 134, 137
Physa subopaca 156
Phycopomatus enigmaticus 102, 103, 105
Pila ovata 138
Pinctada radiata 18, 56
Pirenella conica 53, 74, 96, 102
Pisces 10, 11, 13, 15, 29, 37, 49, 53, 59,
    61, 62, 74, 78, 80, 81, 82, 84, 97, 102,
    111, 129, 120, 126, 127, 128, 133, 134,
    135, 136, 137, 138, 141, 142, 146, 147,
    150, 151, 152, 153, 155, 161, 163
Pisidium 142, 148, 150
Pisidium annandalei 148
Pisidium casertanum 148, 154
Pisidium personatum 148
Pisidium subtruncatum 148
planarians 111, 126
Plecoptera 111, 142, 145, 146, 150, 155
Plesionika 49
Plumatella 158
Podocnemis 5
Polychaeta 61, 102
Polyclinidae (Ascidiacea) 49
Polypedilum aegyptium 151
Pontellina plumata 53
Porites 11, 12
Portunus hastatus 54
Posidonia oceanica 49
Pomacentridae (Pisces) 81, 84
Pontostratiotes 83
Potamogeton 163
Potamogeton pectinatus 102, 103, 125
Potamon potamios 19, 155
Potamon p. ghab 131, 134
Potamon p.palaestinensis 131

Potamon p.setiger 129, 133
Potamonectes cerisyi 139
Potamonectes lanceolatus 139
Potamothrix bavaricus 123
Potamothrix heuscheri 123
Potomida littoralis semirugata 133
prawns 6, 19, 84, 102, 120, 121, 127, 130, 163
Proasellus coxalis 121, 154, 155, 159
Processa canaliculata 49
Procladius choraeus 125
Prosopistoma 6
Protozoa 116
Psammoryctides barbatus 142
Pseudagrion 139, 140
Pseudagrion syriacus 151
Pseudagrion t. hulae 140, 152
Pseudamnicola solitaria 120, 126
Pseudobradya barroisi 125
Pseudochromidae 82
Pseudocyclops 94
Pseudunio homsensis 133
Pteropoda 29, 52, 53, 78
Pterosagitta draco 78
puffer fishes 59
Purpura haemastoma 53
Pyramidaellidae (Gastropoda) 57
Pyrosoma (Tunicata) 52

Rachycentrum canadum 59
Radix peregra 131
rainbow trout 163
Rana ridibunda 159
Ranatra pavipes vicina 140
red algae 54
Rhabdocoela 156
Rhincalanus nasutus 78
Rhinoclavis(=Cerithium) kochi 57, 61, 81
Rhizophora mucronata 75, 77
Rhizopoda 116
Rhyothemis semihialina syriacum 139, 152
Ridgewaya typica 95
Rissoina bertholetti 57
Robertsonia salsa 96
Rotifera 111
Ruppia maritima 102

Sabella pavonina 51
Sagitta enflata 78
Sagitta neglecta 59
Sagitta pacifica 78
Sagitta serratodentata 52
Salamandra salamandra 159
Salaria fluviatilis (=Blennius fluviatilis) 126

Salmo gairdneri 163
Salmoneus jarli 53
Salpidae (Thaliacea) 78
Sardinella aurita 63, 64
Sargassum dentifolium 77
Sargocentrum cf. rubrum 12
Sarotherodon aureus 126, 163
Sarotherodon galileus 126
Sarotherodon niloticus 163
Sarotherodon mossambicus 163
Schizaster canaliferus 51
Sciaena aquila 74
Scleractinia 27, 50, 82, 86, 87, 89, 90, 97, 99
Scorpaenidae (Pisces) 81
sea urchins 50, 71
Semisalsa 125
Semisalsa contempta 131
Semisalsa longiscata 120
Sergestidae (Decapoda) 52
Serpulidae (Polychaeta) 50, 59, 94, 103, 105
Serranus 53
Sepiola 49
sharks 59, 84
shrimps 93
Siderastrea 82, 89, 97
Siganus rivulatus 62
Sigara 150
Sigara lateralis 155
Sigara marginata 155
Silhouetta aegyptica 59
Sillago sihama 59
Sillaginidae (Pisces) 59
Siluroidea (Pisces) 138
Silurus glanis 135
Simuliidae (Diptera) 142, 146
Siphonophora 52
Sipuncula 49
Solitariopagurus profundus 84
Sparus aurata 102
Sphaeriidae (Bivalvia) 141, 142
Sphaeroides spadiceus 58
Sphaeriodiscus placenta 74
Sphaeroma hookeri 107
Sphaerozius nitidus 59
spiny lobsters 49
Spirobranchus tetraceros 59
sponges 49, 50, 51, 94, 125
Stephanolepis diaspros 58
Streptocephalus torvicornis 156
Strombidae (Gatropoda) 81
Strombus bubonius 21, 22, 53
Strombus fasciatus 81
Stylodrylus lemani 142
Stylophora 11, 29, 82, 89, 97

Stylophora pistillata 90
Sympetrum decoloratum 139
Syncarida (Crustacea) 6
Syngnathus 11

tadpole shrimps 158
Tanypus punctipennis 125
Tardigrada 111, 141
Terebra 81
Tetrosomus gibbosus 59
Thais carinifera 57
Thais haemostoma 22
Thalassia hemprichi 99
Thaliacea 52
Theodoxus 120, 126
Theodoxus jordani 125, 131, 147
Thermocyclops galebi 102
Thermosbaenacea (Crustacea) 6, 51
Tilapia 102, 119, 135, 163
Tilapia zillii 126, 138, 151, 163
Tilapiinae (Cichlidae) 151
Tor canis 128
Torquineger flavomaculatus 59
Trichoptera 111, 120, 140, 145, 146, 150, 153, 155
Tricladida 120, 142, 146, 150
Trionyx triunguis 137, 138
Triops sp. 158
Triops cancriformis 156, 158
Tripneustes gratilla 79, 97
Tripterygiidae (Pisces) 82
Tristramella sacra 126, 129, 138
Tristramella simonis intermedia 126, 152
Tristramella s. simonis 126, 138
tritons (Gastropoda) 49
Triturus vittatus 158, 159
Tropocyclops confinis 151
Tubificidae (Oligochaeta) 123
Turbellaria 79
Turbinaria elatensis 82
Typhlocaris galilea, 6, 7, 129
Typhlocirolana reichi 6, 129
Typhlocirolana steinitzi 6, 7

Uca inversa 76
Uca tetragonon 76
Uca urvillei 76
Umbrina cirrosa 74
Unio crassus damascensis 129
Unio mancus 140
Unio terminalis delicatus 131, 133
Unio t.terminalis 131
Unio tigridis 140
Unionidae 140
Uranotenia unguiculata 155
Upeneus asymmetricus 61

Upeneus moluccensis 61
Urothemis edwardsi hulae 139

Valvata saulcyi 147, 156
Veliadae (Hemiptera) 140
Vermetidae (Gastropoda) 54
Vermetus triqueter 54
Vermiliopsis pygidilis 94
Viaderiana meseda 84

Victoriella pavida 105
Vinciguerra mabahiss 84

water hyacinth 138
water mites 111, 141

Xanthidae (Decapoda) 59

Zebrasoma xanthurum 82
Zeus faber 12
Zostera 49

If you have any concerns about our products,
you can contact us on
ProductSafety@springernature.com

In case Publisher is established outside the EU,
the EU authorized representative is:
**Springer Nature Customer Service Center GmbH
Europaplatz 3, 69115 Heidelberg, Germany**

Printed by Libri Plureos GmbH
in Hamburg, Germany